2.95

HOW TO GROW FLOWERS FROM SEEDS

Elvin McDonald

D0921783

 VAN NOSTRAND REINHOLD COMPANY
New York Cincinnati Toronto London Melbourne

ACKNOWLEDGMENTS

Special thanks to
Jacqueline Heriteau,
Bill Mulligan,
and Mary Otto.

Copyright © 1979 by Litton Educational Publishing, Inc.
Library of Congress Catalog Card Number 77-19101
ISBN 0-442-80570-5   cloth
ISBN 0-442-80594-2   paper

Printed in the United States of America

Published in 1979 by Van Nostrand Reinhold Company
A division of Litton Educational Publishing, Inc.
135 West 50th Street, New York, NY 10020, U.S.A.

Van Nostrand Reinhold Limited
1410 Birchmount Road
Scarborough, Ontario MlP 2E7, Canada

Van Nostrand Reinhold Australia Pty. Ltd.
17 Queen Street
Mitcham, Victoria 3132, Australia

Van Nostrand Reinhold Company Limited
Molly Millars Lane
Wokingham, Berkshire, England

16 15 14 13 12 11 10 9 8 7 6 5 4 3 2 1

Library of Congress Cataloging in Publication Data

McDonald, Elvin.
  How to grow flowers from seeds.

  Includes index.
    1. Flower gardening.   2. Seeds.   3. Plant
propagation.   I.   Title.
SB406.83.M3 1978   635.9´42   77-19101
ISBN 0-442-80570-5
ISBN 0-442-80594-2 pbk.

Contents

Introduction

Helping Nature Work
Miracles

Growing flowers from seeds has been a passion of mine since I was about three years old. If the old saying is true, that the best gardeners are those who succeed despite monumental environmental problems, then I must have been a gardener's gardener by the time I reached my early teens. The western, panhandle part of Oklahoma where I grew up is often either bitterly cold, dry, and windy or beastly hot, dry, and windy—in other words, it is no Eden. Yet, by starting the most delicate flower seeds indoors, and giving the others protected seedframes outdoors, I was able to grow a garden that was seldom without flowers from earliest spring until fall frost.

Later, in my twenties, I grew gardens in two far more hospitable climates—first in Kansas City, Missouri; second in Levittown, Long Island (New York). In time, as I became friendly with other neighborhood gardeners, I discovered that the old-time locals in both states thought they were gardening under great environmental disadvantages and, to some extent, I suppose they were right. Wherever we live, nature often seems to be at cross-purposes with our own, especially where gardening is concerned. Now that I do my outdoor flower raising in northwestern Connecticut, I've discovered that another part of the world, which I thought was a real garden spot, has its own set of problems and limitations.

What I've decided as a result of my gardening experiences is that a major satisfaction of growing a flower garden—wherever you live—is the sense of helping nature work its miracles. Failure is sometimes attributable to human error, sometimes to a whim of nature. In any event, the only way to learn to be a really good gardener is to dig in and not be afraid of making mistakes. Even though it is gratifying to have seeds sprout on schedule, seedlings grow vigorously, and a space of bare ground turn miraculously into a flower garden, there is not much to be learned. But when something goes wrong— that's when we have the opportunity to learn one or more basic gardening lessons.

Probably the easiest way to grow a flower garden is to purchase all started seedlings or established divisions. If you can afford it, fine. But growing from seeds offers more than a savings in money. First, it is an enormously satisfying experience. Some quick annuals bloom in two months, there are perennials that may not have the first flowers for several years, and in between there are literally hundreds of beautiful blooms you can grow in a matter of one season.

Second, from the moment you plant flower seeds, or any other kind for that matter, you have something to look forward to—a pyramiding experience that begins with the waiting and watching for the first green sprouts of seedlings and that climaxes with the flower buds unfurling their petals into glorious blooms.

And finally, it is simply not possible to purchase started plants or established divisions of some of the world's finest flowers. Seeds are the answer and most are readily available from the catalogs of specialists (see Appendix). Others can be collected from your own garden, that of a friend, or possibly from the wilds. And some of the specialized plant societies (see Appendix) have seed exchanges that often offer extremely rare species.

Although not every flower in the world will be found in the pages that follow, most of the know-how for starting almost any kind of flower seed is included. If you want to grow something that is not in these pages, start with this basic set of rules: sow in the spring; use a sterile planting medium such as pasteurized packaged potting soil, vermiculite, or milled

Flower seeds and ornamental grasses may be purchased from local racks or ordered by mail from specialists.

sphagnum moss; cover the seeds to the depth of their own thickness; enclose the moistened planting in plastic; provide average temperatures of 70° F.; place in bright light, natural or fluorescent, but not in direct sun; when seedlings show, remove the plastic, and provide about a half-day of direct sun or grow six inches directly below two 20- or 40-watt fluorescents burned fifteen to sixteen hours out of every twenty-four; do not let the growing medium dry out at any time; provide fresh air ventilation.

Now, if this routine fails with a seed, you can begin trying other approaches. Perhaps it needs continual light or darkness to effect the most successful germination; or freezing in the starting container for two to four months before placing in warmth; or planting outdoors in early spring while the soil is still cool; or maybe the seeds have a hard coat or shell and need to be soaked in water at room temperature for twenty-four hours before planting. All these techniques and more are explained in the pages that follow.

How to Start Flower Seeds Indoors

Although it is possible to grow a garden filled with flowers simply by sowing the seeds directly in the ground outdoors, as described in Chapter 2, there are numerous advantages to starting certain kinds indoors. Some plants have tiny seeds that have little chance of survival unless they are started indoors where you can control light, temperature, and moisture. Others need a longer growing season than the local climate affords. And finally, at the end of a long winter, sowing some seeds indoors can be a great morale booster.

Equipment: Anything three or four inches deep that will hold soil can be a container for sowing seeds indoors. Garden centers and catalogs offer flats (shallow rectangular boxes) and pots of various sizes and kinds. Flats are most practical for the first stage of growth. There are Jiffy Sevens; flats with individual compartments; flats made of pressed cardboard, pressed peat moss, plastic, or wood. There are also plastic mesh trays meant to hold individual, small peat pots.

The pressed peat containers, and some pressed cardboard containers, have a real advantage: they can be planted with the seedlings right in them. The peat will disintegrate, adding organic, humuslike materials to the soil. Pressed peat is a particular advantage for flower seedings that react badly to transplanting—nasturtium and morning-glory, for instance.

The plastic and wooden flats or pots are durable—that's their advantage. You'll be able to use them next year, while the peat and pressed cardboard containers are gone.

Supplies for starting flower seeds indoors: wooden seed flat, seeds, plastic pot, peat pots, watering can, plastic bag to enclose planting, and potting soil. Courtesy of the United States Department of Agriculture.

If you aren't ready yet for investment in special materials for starting seeds indoors, raid the kitchen. One- and two-quart waxed milk cartons, sliced down the center to make long narrow boxes, are fine planters; square plastic gallon containers, sliced down the center (the half spout or bottle neck can act as a drain to carry off excess water), are also good. Cans make fine pots of various sizes for larger seedlings, and two-pound coffee tins make serviceable big pots for large seedlings. You can also use discarded kitchen equipment—glass refrigerator trays and cracked or chipped bake trays. Old cookie sheets make fine saucers for groups of little pots. If you have any to spare, great big saucers from large houseplant pots make excellent flats for starting seeds.

Seed Starting Mediums: Commercial mediums for starting seeds include vermiculite, perlite, milled or screened sphagnum moss, and sphagnum peat moss. These are sterile, light, and porous mediums—three big advantages. Bagged potting soil for indoor plants is pasteurized and is also suitable for starting seedlings.

If you don't want to use one of these commercial mediums, you can make your own soil by combining one-third garden loam, one-third sharp sand (builder's sand), and one-third fine ground sphagnum peat moss or any of the commercial mediums mentioned above.

To get fancy, mix together one part sphagnum peat moss, two parts garden soil or potting soil, two parts sand or vermiculite, and one part dried cattle or sheep manure—preferably sheep. For each bushel of this mixture, add nine level tablespoons of superphosphate, nine level tablespoons cottonseed meal, four level tablespoons sulfate of potash, and two level tablespoons ground limestone.

If you use garden soil or garden sand for any of these mixtures, it is best to pasteurize it. "Damping off," a moldy fungus, is a hazard to seedlings, and it, along with some harmful nematodes, occurs in unpasteurized soil. Most practical for pasteurizing large quantities of soil is an oven or a

Growing mediums for starting flower seeds: back row, left to right, perlite, sphagnum peat moss, milled (screened) sphagnum moss; front row, left to right, vermiculite, sand, and soilless mix.

barbecue heated to 180° F. Place the soil in a very large kettle, add one cup of water for each four quarts (sixteen cups) of soil, and bake forty-five minutes. Dump the soil onto several sheets of newspaper to cool, and let stand twenty-four hours before mixing it in with the other ingredients.

Natural and Fluorescent Light: Light suitable for starting seedlings can be any sunny window or fluorescent setup. If you plan to start a few seedlings, you probably have enough sunny window space. But if you want to start everything, or almost everything you would otherwise buy as started seedlings, consider acquiring a three-tiered fluorescent plant stand. One of the best has three two-by-four foot shelves, each lighted by two 40-watt fluorescent tubes. You can also make your own fluorescent grow shelves: install two 40-watt fluorescent tubes, one Warm White and one Cool White, side by side under a reflector and about fifteen inches above a table or shelf top.

Seedlings growing under fluorescent installations can be placed anywhere. A cool basement (not damp), with some moisture in air that circulates freely, is a great place to start most flower seeds. A closet will do, too, as long as there is good air circulation; stuffy hot corners are not good places to start seedlings (or for growing much else either).

Fixtures come complete with instructions on the distance soil and plants should be from the light and the number of hours the lights should be on. If you have no guide, try placing plants so that seedling leaves will be about six inches directly below the tubes, and start seeds at fifteen to sixteen hours of light daily. If the seedlings show sparsely placed leaves on spindly stalks, chances are there's not enough light, or they aren't close enough to the light. If seedlings are low-growing, with distorted leaves, there may be too much light, or the light may be too close.

When to Start Seedlings: Although recommended dates are given for specific plants throughout this book, as a general

A fluorescent-light garden is an ideal place to start most flower seeds. Use two 20- or 40-watt tubes in a reflector. The leaves of the seedlings should be about six inches directly beneath the tubes. Burn the lights fifteen to sixteen hours out of every twenty-four.

A three-tiered fluorescent-light garden like this one provides approximately twenty-four square feet of growing space. Grow seeds that need cooler temperatures on the bottom shelf, those that need more warmth on the top.

rule, sow seeds indoors six to ten weeks before they are to be set out in your area. The phrase "planting-out time" used with individual flowers included in this book means at or around the average date of the last killing frost. If you don't know this date for where you live, ask a neighbor who gardens successfully, someone who seems knowledgeable at your local nursery or garden center, or call the local office of the U.S. Weather Service.

How to Plant Seeds: The rule of thumb governing flower seed plantings is to plant at about three times the depth of the height or diameter of the seed itself. If seeds are dust-fine, they are not covered.

In the growing guides for individual annual, biennial, perennial, and bulb flowers (Chapters 3, 4, 5, and 6), you will notice that some need light in order to sprout, others need continual darkness. If you are planting seeds that need light, do not cover them with any medium; simply press the seeds into the surface with your fingers or the bottom of a drinking glass. If the seeds need continual darkness, cover them with planting medium if they are large enough; if they are tiny seeds, do not cover, but keep the planting in a dark closet or cupboard until sprouting begins, and then bring it to light.

Fill containers two-thirds full of potting soil, or whatever growing medium you have selected. Moisten the soil by placing the container in a basin of tepid water until beads of moisture show on the surface. Remove and allow to drain. The soil should be moist but not soggy. With a sharp pencil draw small furrows one-quarter inch deep along the soil surface, and sow large seeds in the bottoms of the furrows. Before sowing tiny seeds, fill the furrows with vermiculite and sow in the vermiculite (or drop tiny seeds on the soil surface). Sow larger seeds an inch or so apart. Sow tiny seeds one-quarter to one-half inch apart. Cover the seeds with a thin layer of vermiculite, milled sphagnum moss, or potting soil (except as noted in the preceding paragraph). Place a plastic tent (a cleaner's pliofilm bag for instance) over the flat to keep

Fill container with half-and-half mixture of packaged potting soil and vermiculite. Courtesy of the United States Department of Agriculture.

Add a half-inch layer of sterile medium such as vermiculite, which is shown. Courtesy of the United States Department of Agriculture.

Moisten well (here, with the fine rose attachment on a watering can) before planting seeds. Courtesy of the United States Department of Agriculture.

After seeds are sown, press them in place with fingers, palm of hand, or block of wood. Place the container in a basin of water; when beads of moisture show on the surface, remove, allow to drain, and then enclose the planting in a plastic bag (or cover with glass). Courtesy of the United States Department of Agriculture.

If air temperatures are too cool to suit the sprouting needs of a particular kind of seed, use an electric soil-heating cable, as shown in this flat. Courtesy of Gro-Quick.

moisture in, and put the flat in a room with temperatures between 65° F. and 75° F. (except as noted for individual plants that require cooler or warmer temperatures for optimum germination). The tenting should keep the soil damp without further watering until the seedlings are up. However, if the soil surface appears at all dry, very gently water the soil again with a fine spray, or set the container in a basin or sink to absorb water from below as previously described.

During the germination period, if there are any signs of damping off—a cottonlike white fungus growing along the soil surface—remove the tenting, and air the plantings well. Damping off usually is a result of too much soil moisture and lack of air circulation.

Care After Germination: Once the seedlings are up, remove the tenting, and begin a careful watering program—keeping the soil evenly moist, never soggy, but never letting it dry.

At this stage seedlings of plants that grow best in warm weather can go into a room where temperatures are around 70° to 75° F. Hardy annuals such as snapdragon, and perennials like primroses, that do best in cool weather will do better if temperatures are between 60° and 70° F. If the perfect climate isn't available, all will probably manage at temperatures between 65° and 72° F.

Once seedlings show two pairs of true leaves, it is time to consider thinning. With a knife tip, dig out the weakest in crowded groups, elevating the strongest to stand free with room to grow. As the seedlings develop, continue to thin. If you find yourself removing sturdy little seedlings toward the end of this thinning process, transplant them to new flats. Dig a planting hole for each seedling, lift it gently by the stem, set it in the hole, and very gently press a little soil up around it until it stands straight. Those that don't make it—discard. Your conscience will have been eased by the effort made.

Transplanting to Larger Containers: Small plants may stay in the original planting flat, if the flats don't become so crowded that they inhibit growth. However, larger plants often benefit from transplanting to larger containers for the second stage of indoor growth.

Prepare the second-stage flats or pots by filling them two-thirds full with a half-and-half mixture of potting soil—as described previously—and vermiculite or perlite. Vermiculite and perlite make the soil very light so that tiny roots grow well.

At this stage, seedling roots are usually small enough to be lifted whole from the original growing medium into small holes dug with a spoon in the new potting soil. Repeat the procedure described above for moistening the soil, plant the seedlings, and press the soil gently around the roots until stems stand upright.

Some gardeners transplant the big plants a third time before their final trip to the outdoor garden. The procedure is the same as when transplanting to a second container. But this time, choose the biggest container you have available—for instance, a gallon plastic bottle with the top cut off, a two-pound coffee tin, or a six- to ten-inch pot—so the plant has all the room it needs to grow. However delayed planting-out weather may be, space will encourage plants to maximize growth in the weeks left before planting time.

Hardening Off: One week before seedlings are due for planting in the garden, move them to a protected warm spot outdoors—for instance, a sheltered corner of the house where there is some sun part of the day, or a porch or patio that is sunny for a few hours. Each day, move them into a slightly less sheltered spot that gives them more sun and more breeze. This prepares the seedlings for the final trip to the open garden. Check moisture daily; they dry out much more quickly in the open. Water as needed to keep the soil evenly moist.

Transplanting Outdoors: Before seedlings can be transplanted, soil must be readied (as described in Chapter 2).

Transplanting seedlings from starting flat into individual peat pots filled with soilless growing medium. Position the seedlings at the same level in the soil as they grew before. Courtesy of the United States Department of Agriculture.

Before planting seedlings directly in the garden after they have been started early indoors, condition them to the open air gradually —a process called "hardening off." First place seedlings on a porch or in the shade of low-branched shrubs for a few days, then transplant to the open garden.

When taking the seedlings to the garden, bring a plastic bucket filled with lukewarm water—water that has been standing in the sun a few hours—in which you have dissolved starter solution, a liquid vitamin-hormone stimulant, or a liquid fertilizer (available at garden centers). Follow container directions for quantities to add. You will find a large paper cup, a small sharp kitchen knife, and a trowel to be helpful.

With a row or other areas marked, dig planting holes for the seedlings with the trowel. The holes should be spaced wide enough apart so mature plants have plenty of room to grow; holes should be a bit deeper and larger than the roots of the seedlings.

If the seedlings are in flats, cut them apart (cutting is less damaging to roots than tearing). If plant seedlings are in individual pots, rap the pot bottom sharply on a rock or hit the rim with the end of the trowel; spread your fingers across the top of the pot; turn the pot upside down, and the plant will fall into your fingers. If it doesn't slide out easily, loosen the soil around the rim gently with the knife and repeat the maneuver.

Pour a little water from the pail into the planting hole, then, with the rootball (peat or cardboard pot) cupped in your hand, dip the plant quickly into the planting solution in the bucket, and at once place it upright in the planting hole. Push soil in around the rootball halfway to the top of the hole. Add another cupful of water, then fill the hole with more soil, and firm the rootball very gently into place. Make a small depression around the base of the stem, and add a bit more water. Tug gently on the plant stem. If it resists uprooting, it is properly planted.

The best time to transplant is toward the end of the day, after the heat of the sun is gone. When all the planting is finished, water everything gently with a fine spray.

Protection for Newly Set-Out Seedlings: If plants have not been hardened off, and must be planted for some reason, protect them the first few days outdoors with big brown paper bags from the supermarket. If setting seedlings out early, before the weather has warmed reliably, protect them with the plastic "hot caps" (little tents sold by garden centers and catalogs) used by professionals who want a headstart on warm weather.

How to Start Flower Seeds Outdoors

There are essentially two ways to start flower seeds outdoors—either in protected frames or directly in the ground where they are to grow and bloom. In either case, the first requirement is that the soil be well prepared.

Preparing the Soil: To get good bloom from flowering plants, the soil must have three properties: good structure, high fertility, and a correct pH balance.

Structure refers to the composition of the soil. The good garden loam everyone talks about is usually a combination of soil, sand, and humus in approximately equal parts. Soil contains nutrients; sand allows air and water to penetrate the soil; and humus retains moisture.

Fertility has to do with plant food. Many of the nutrients plants require are already part of the soil. Others you will add. However, unless the soil is in correct pH balance, nutrients get locked up and aren't available to the plants.

To test your soil, on a day when the ground is damp (not soggy), pack a handful of soil as though you were making a snowball. If the soil won't stick together in a ball, chances are it has too much sand. If it quickly makes a ball, but the ball won't crumble under slight pressure from your thumb, chances are there is too much clay. If it balls easily, crumbles easily under thumb pressure, chances are it is that miraculous stuff we call good garden loam.

If the content is too sandy, add humus and topsoil. Humus can be, ideally, compost. However, if no compost is available, woodlands are full of soil rich in humus; so are garden centers, by the bag or by the yard (roughly, a small truckload). If topsoil is needed, search out a supplier who has a reputation for having superb topsoil, and pay whatever he or she wants for it. Buy less if you must, but don't import topsoil with a shady reputation. It may be full of weeds (weeds much worse than any your own yard boasts), or rocks may make up much of its weight. If your soil has too much clay, add sand and humus. The sand to be used must be the kind called "sharp" or builder's sand. Inert fine sand and sea sand won't do.

How Much to Add: The local Agricultural Extension Service at your state university will test a soil sample and tell you exactly what to add. You can also figure it out yourself. Cover a test patch of damp (not wet) soil about two feet square with a one-inch layer of whatever you think is missing. Mix the additive thoroughly with the soil to a depth of eighteen inches and try the snowball test again. If it still doesn't work, keep adding the missing element(s) an inch at a time until the soil will ball properly. Keep a record of what you have added, and use it as a guide for correcting the composition of the garden soil.

Fertilizing a new garden area requires the same careful approach. You can buy whatever the local garden center recommends to other flower gardeners in the area, and probably it will be exactly right, since they have been dealing with other local gardeners' problems for some time. You can also send a plug of soil to the state Agricultural Extension Service, and they'll test it and make recommendations. If you are so inclined, you can also buy a soil-testing kit, follow directions, and investigate for yourself. It's rather fun and fairly accurate if you pay attention to the instructions.

Both the Agricultural Extension Service and soil-testing kits are likely to recommend a complete fertilizer, one containing nitrogen, phosphorus, and potash. The numbers on bags of complete fertilizer refer to these three basic elements, in that order. A common recommendation for flower gardens is

5-10-5. A common recommendation for quantity is to dig complete fertilizer into the soil at the rate of fifty pounds for every 2,500 square feet of garden surface.

Once the soil shows a suitable fertility level on your soil-testing kit, you might maintain that level by annually adding a bag or two of dried sheep or cow manure to the soil. These are organic fertilizers, which I prefer to use instead of the chemical types. They should be dug into the soil as soon as the ground is dry enough for gardening in early spring. Each fall, you might also add a dressing of organic phosphate or bone meal. Most of my gardens have benefited from application of either or both. Bone meal contains nutrients valuable to plants and is a slow-release fertilizer that lasts a long time.

As plants grow and deplete the soil, soil becomes acidic. That's why we lime lawns periodically. Flower gardens benefit from occasional liming, too. Here again the Agricultural Extension Service, or your own soil-testing kit, can provide some answers. Ideal soils for most flowers are between pH 6.0 and pH 6.8. Most American soils are in this range. However, if yours is not, the acidity, or lack of it, can lock up nutrients vital to your plants. Ground dolomitic limestone is the additive that sweetens soil, and ammonium sulfate is one of the additives proposed when soil is "too sweet," in other words, too alkaline. But this is less common than soil that is too acidic.

When adding either of these elements to your soil—or any other elements for that matter—follow instructions given on containers. Less is better than too much.

Digging a New Garden: If your soil has never been turned before, begin work during the fall prior to the season in which you intend to plant; if this is not possible, start as soon as the soil can be worked in the spring. Strip away the sod on top (grassy growth and its roots), and turn the soil to a depth of about eighteen to twenty inches down. Break soil clods before returning them to the earth. The ground must be crumbly and loose before you can plant. I use a digging fork for this operation—and a lot of elbow grease.

A rototiller is a blessing, one worth paying money for when the garden needs to be dug brand new. A rotary tiller digs and fluffs the soil. But don't use one until you've removed the sod; there may be weeds that propagate by means of underground runners—quackgrass is one—and once chopped by a tiller, they will propagate like crazy.

After the soil has been dug, or tilled, it is time to dig in additives—manures, plant foods, acidifiers or sweeteners. If you are liming, apply before fertilizing. Leave a few days between the stages if there's time. If you have precious compost to add to the garden, dig it in only when and where you will actually plant (unless you have a real stockpile).

How to Sow Seeds: Depending on the plan of your garden, flower seeds can be planted in rows (as in a vegetable patch), in drills placed in irregularly shaped patches within a planting bed, or scattered in an area just as if nature might have strewn them there.

Size of seed determines to what depth you cover it with soil. Dust-size seeds—e.g., those of *semperflorens*, or wax begonias—should be started indoors under controlled conditions. But slightly larger, hardier kinds—like *Petunia*, poppy, and rose moss—fare quite well when sown in the garden, provided the soil is well-prepared and the area is kept moist until the seedlings are well along. Small seeds, like the ones just mentioned, hardly need any covering at all in the garden—just pat them into the soil surface with the palm of your hand, or walk over the area. If you've broadcast them in a semi-wild area, plant extra seeds and let nature take its course. Larger seeds—those of nasturtium, marigold and *Zinnia*, for example—do nicely with one-quarter- to one-half-inch covering, or a coverage equal to two or three times their diameter.

After seedlings are making healthy growth in the garden, you have to determine if thinning is needed. You can follow instructions given on the seed packet, or use common sense. For example, if seedlings seem to be growing weak and stretching for light, thinning is probably needed. Another rule of thumb is to think about the approximate size of a plant at maturity. *Petunias* that will grow about twelve inches tall will do nicely spaced twelve inches apart. Marigolds two feet tall might have eighteen-inch to two-foot spacing. You can use seedlings thinned from one area to "patch" part of the row or bed that did not germinate well; simply transplant them to another part of the garden—or give them to a neighbor.

Coldframes, Hotbeds, and Seedframes: A hotbed or coldframe is the ideal place for giving flower seeds an early start in late winter or early spring and for wintering-over biennial and perennial seedlings before setting them into permanent growing positions in the garden at the beginning of the second year's gardening season. When the glass or plastic used to cover a hotbed or coldframe is removed at the onset of warm, frost-free weather, the area inside can be used as a seedframe for starting biennials, perennials, and bulbs (discussed in Chapters 4, 5, and 6).

A coldframe is a relatively simple structure that extends the growing season into early spring or very late fall. It is particularly useful for starting slow-to-germinate perennials. Seeds can go into a coldframe several weeks, four to six at least, before they can go into the open ground. The glass top for a coldframe, referred to as a "sash," is carried by hardware stores and garden centers. The standard size is three-by-six feet. There are smaller sashes, two-by-four feet, also. The base can be made of poured concrete, concrete blocks or brick. It can also be made of moisture-resistant woods such as redwood or cypress, or of ordinary wood treated with wood preservatives and painted.

Though glass is desirable—because it lets more infrared rays through than plastic and because it is durable—you can make an inexpensive sash to cover a wood frame with any of the transparent plastic sheets.

If you want to have more than three-by-six feet of coldframe, make a long base four feet wide from front to back, and cover it with as many sashes as you like, laid side by side.

The soil inside the frame must be dug well (as should all garden soil) to a depth of six inches, and then it should be supplied well with humus and nutrients. Beneath that six inches of soil should be a layer of sand, one of straw, and one of gravel.

A coldframe is planted much as the garden is planted, except that seeds go much closer together. During the day, especially when it is warm, the sash should be raised four to six inches, to let in air and to let out excess moisture and heat. Water the coldframe when the soil surface shows signs of becoming dry. Water in the morning; watering at night sometimes encourages damping off.

A hotbed is much like a coldframe, but it is artificially heated. Years ago, it was heated by laying a foundation under the soil of fresh horse manure, which decomposed and gave off considerable heat. Today, horses are scarce and a heating coil and thermostat are used instead. For a beginner in the world of flower gardening, a coldframe offers many of the opportunities a hotbed offers and is much less complex to construct and maintain.

A seedframe can be a hotbed or a coldframe without sash, as explained before, or it can be simply a four-sided wood frame such as a fruit lug without a top or bottom. This should be situated in the garden where it will be protected from stormy winds and rains, as well as hottest sun during the middle part of the day. If light tree shade is not available, use lightweight wood laths spaced about one inch apart on a frame. For best light distribution, place the laths so that they run north and south rather than east and west.

Watering: Flower gardens need watering, a good soaking, about once a week. If the sky doesn't do it, you must. There are many watering devices on the market. The soakers are the best, since they keep leaves dry, and do little to encourage pests or diseases. Next best is one of the tall, rotating arms that are popular with nurseries. Watering in the early morning is the most beneficial.

Weeds and Mulching: Mulch (compost, peat moss, or buckwheat hulls are some of the preferred organic mulches) is great for keeping weeds down, but many gardeners prefer to apply it after the seedlings are up and thriving. Mulch keeps heat from the soil, as well as moisture in the soil; therefore, when the weather is still cool, mulch may hold seedlings back by keeping heat from them. The first days and the first few weeks after planting, weeding is easily done by walking down the garden row with a rake or a hoe and disturbing the soil just enough to uproot those tiny beginning weeds. If you let them get a few inches higher, however, it stops being easy. You have to cut them off with a hoe (as farmers do) or dig them up with a hoe as I do (to the detriment of my arm and shoulder muscles).

Once the seedlings are prospering, a mulch several inches deep keeps the weeds at bay, holds moisture in the soil, and is beneficial in several other ways. It keeps weed seeds from landing on the soil, for instance. The underside of the mulch decays, adding humus to next year's garden. Mulch also keeps leaves and flowers free of mud splash during rainy periods.

Annual Flowers to Grow from Seeds

Annuals are the flowers that grow from seed to bloom, form a new seed crop, and die—all within a single season. They offer not only the quickest, but also the least expensive way to fill a garden with color and fragrance.

Not all of the plants included in this chapter are annuals in the strictest sense. Many of them are tender perennials (geraniums and *Impatiens*, for instance) that grow from seed to bloom in a single season; yet, they will live year after year provided they are not subjected to freezing temperatures.

ABUTILON (flowering-maple)

When to plant: For indoors, plant in any season. For flowering-sized seedlings to plant outdoors, sow seeds 12 weeks before frost-free weather.

Temperature: 70° F.

Special treatment: None required. An easy plant to grow from seeds.

Days to sprout: 20.

Light for seedlings: Half-day or more of sun, or grow in a fluorescent-light garden (15–16 hours of light daily) until the plants start to get too tall. Frequent pinching of the new growth will encourage compact, well-branched plants.

Maturity time: First blooms in 4 to 6 months.

Comments: *Abutilon* is not a true maple, although the leaves are often maple-shaped, but a relative of the hollyhock (Mallow family). The bell-shaped flowers come in many beautiful pastels. If given full sun in winter, with temperatures ranging somewhere between 55° and 70° F., and an atmosphere that is moist with freely circulating fresh air, flowering-maple will be almost everblooming. Outdoors in warm, frost-free weather, this makes a beautiful flowering plant for container gardens of all kinds, including large pots, tubs, window boxes, and hanging baskets. The seeds are listed by Butcher, Park, and Thompson & Morgan (*see also* Appendix).

ACANTHUS (bear's-breech)

When to plant: 8 to 12 weeks before frost-free weather in the spring.
Temperature: 55°–65° F.
Special treatment: None required.
Days to sprout: 20.
Light for seedlings: Half-day or more of sun, or start in a fluorescent-light garden (15–16 hours of light daily).
Maturity time: About 6 months to first flowers.
Comments: *Acanthus* is easy to grow from seeds. In climates where winter temperatures stay generally above 20° F., *Acanthus* may be treated as a hardy perennial. Elsewhere it may be cultivated as an annual or grown in large pots or tubs kept indoors in cold weather. Seeds are listed by Butcher, Park, and Thompson & Morgan (*see also* Appendix).

AFRICAN DAISY: *See Arctotis*

AFRICAN DAISY: *See Gerbera*

AGERATUM (flossflower)

When to plant: Indoors, 8 to 12 weeks before planting-out time in the spring; outdoors, as soon as the soil is warm and there is no danger of frost.
Temperature: 70°–75° F.

Ageratum *seedlings about ten weeks old and ready to transplant to the garden.*

Special treatment: Seeds need light to sprout; sow on surface, and do not cover with planting medium.
Days to sprout: 5.
Light for seedlings: Half-day or more of sun, or start in a fluorescent-light garden (15–16 hours of light daily).
Maturity time: First blooms in about 3 months.
Comments: Flossflower is easy to grow from seeds. Hybrid forms of *Ageratum* are available from most seed companies in varieties with blue, pink, or white flowers.

ALKANET: *See Anchusa*

AMARANTHUS   (Joseph's-coat;   love-lies-bleeding; fountain plant)

When to plant: Indoors, 6 to 8 weeks before planting-out time in the spring; outdoors, as soon as the soil is warm and there is no danger of frost.
Temperature: 70°–75° F.
Special treatment: None required.
Days to sprout: 10.
Light for seedlings: Half-day or more of sun, or start in a fluorescent-light garden (15–16 hours of light daily).
Maturity time: Colorful foliage accent plants within 2 months.
Comments: Easy to grow from seeds, *Amaranthus caudatus* (love-lies-bleeding), *A. tricolor* (Joseph's-coat), and *A. tricolor salicifolius* (fountain plant) are all cultivated for their colored and variegated foliage; the flowers are of little consequence. Seeds are listed by Burpee, Butcher, Park, Suttons, and Thompson & Morgan (*see also* Appendix).

AMMOBIUM (winged everlasting)

When to plant: Indoors, 8 to 12 weeks before planting-out

time in the spring; outdoors, as soon as the soil is warm and there is no danger of frost.
Temperature: 70°–75° F.
Special treatment: None required.
Days to sprout: 5.
Light for seedlings: Half-day or more of sun, or start in a fluorescent-light garden (15–16 hours of light daily).
Maturity time: First blooms in about 5 months.
Comments: Easy to grow from seeds, *Ammobium* grows 2 to 3 feet tall with showy, white flowers that may be air dried for long-lasting winter bouquets. *Ammobium* is variously listed as a hardy annual, biennial, or perennial; if seeds are started early indoors, it should bloom during the first season. Seeds are listed by Butcher, Park, and Thompson & Morgan (*see also* Appendix).

ANAGALLIS (pimpernel)

When to plant: Indoors, 6 to 8 weeks before planting-out time in the spring; outdoors, as soon as the soil is warm and there is no danger of frost.
Temperature: 70°–75° F.
Special treatment: None required.
Days to sprout: 10.
Light for seedlings: Half-day or more of sun, or start in a fluorescent-light garden (15–16 hours of light daily).
Maturity time: First blooms in about 3 months.
Comments: Easy to grow from seeds, pimpernel is excellent for front-of-the-border plantings in sun with sandy, well-drained soil. *Anagallis* seeds are listed by Butcher and Thompson & Morgan (*see also* Appendix).

ANCHUSA (bugloss; alkanet)

When to plant: Indoors, 8 to 12 weeks before planting-out

time in the spring; outdoors, as soon as the soil is warm and there is no danger of frost.

Temperature: 70°–75° F.

Special treatment: None required.

Days to sprout: 10.

Light for seedlings: Half-day or more of sun, or start in a fluorescent-light garden (15–16 hours of light daily).

Maturity time: First blooms in about 4 months.

Comments: Easy to grow from seeds, *Anchusa* behaves variously as a hardy annual, biennial, or perennial, depending on the species or variety and local conditions. It is often called a summer forget-me-not because of the clusters of blue flowers, often with a contrasting eye of white in the center of each. Seeds are listed by Burpee, Butcher, Park, Suttons, and Thompson & Morgan (*see also* Appendix).

ANGEL'S-TRUMPET: *See Datura*

ANIMATED OATS: *See Avena*

ANTIGONON (coralvine; Mexican creeper; chain-of-love)

When to plant: Anytime.

Temperature: 70°–75° F.

Special treatment: None required.

Days to sprout: 20.

Light for seedlings: Half-day or more of sun, or start in a fluorescent-light garden (15–16 hours of light daily).

Maturity time: First flowers in 4 to 5 months.

Comments: In tropical and subtropical climates, this beautiful flowering vine behaves as a perennial; elsewhere it may be started indoors about 12 weeks before warm, frost-free weather is expected outdoors. *Antigonon* also makes a choice flowering vine to cultivate in a moderate-to-warm greenhouse. The seeds are listed by Butcher and Park (*see also* Appendix).

ANTIRRHINUM (snapdragon)

When to plant: Indoors, 6 to 8 weeks before planting-out time in the spring; outdoors, as soon as the soil is workable in the spring.
Temperature: 65° F.
Special treatment: Seeds need light to sprout; sow on surface, and do not cover with planting medium.
Days to sprout: 10.
Light for seedlings: Half-day or more of sun, or start in a fluorescent-light garden (15–16 hours of light daily).
Maturity time: First blooms in about 3 to 4 months.
Comments: Snapdragons are easy to grow from seeds, especially if they are started indoors. Hybrid snapdragons are available from most seed companies (*see also* Appendix).

APHANOSTEPHUS (lazy daisy)

When to plant: Indoors, 6 to 8 weeks before planting-out time in the spring; outdoors, as soon as the soil is warm and there is no danger of frost.
Temperature: 70° F.
Special treatment: None required.
Days to sprout: 8.
Light for seedlings: Half-day or more of sun, or start in a fluorescent-light garden (15–16 hours of light daily).
Maturity time: First blooms in about 3 months.
Comments: *Aphanostephus* is easy to grow from seeds which are listed by Park (*see also* Appendix).

APPLE-OF-PERU: *See Nicandra*

Antirrhinum *(snapdragon) seedlings look like this when they are about five weeks old and ready to transplant to the garden.*

ARCTOTIS (African daisy)

When to plant: Indoors, 6 to 8 weeks before planting-out time in the spring; outdoors, as soon as the soil can be worked and there is little danger of frost.
Temperature: 60°–70° F.
Special treatment: None required.
Days to sprout: 10.
Light for seedlings: Half-day or more of sun, or start in a fluorescent-light garden (15–16 hours of light daily).
Maturity time: First blooms in about 3 months.
Comments: Easy to grow from seeds, *Arctotis* has everblooming daisy flowers in a range of both brilliant and pastel colors; they are excellent in the garden and for cutting. Seeds are listed by Burpee, Butcher, Park, Suttons, and Thompson & Morgan (*see also* Appendix).

ASCLEPIAS FRUTICOSA (Gomphocarpus fruticosus)

When to plant: Indoors, 6 to 8 weeks before planting-out time in the spring; outdoors, as soon as the soil is warm and there is no danger of frost.
Temperature: 70° F.
Special treatment: None required.
Days to sprout: 15.
Light for seedlings: Half-day or more of sun, or start in a fluorescent-light garden (15–16 hours of light daily).
Maturity time: Flowers, followed by bronze to chartreuse yellow fruits, in 5 to 6 months.
Comments: This unusual annual is cultivated for the attractive, colorful fruits which dry easily for long-lasting winter bouquets. The seeds are listed by Park as *Gomphocarpus* (*see also* Appendix).

ASTER, ANNUAL: *See Callistephus*

AVENA (animated oats)

When to plant: Outdoors where they are to grow and mature as soon as the soil is warm and there is no danger of frost.
Temperature: 70°–75° F.
Special treatment: None required.
Days to sprout: 5.
Light for seedlings: Half-day or more of sun.
Maturity time: Panicles of oatlike florets in 3 or 4 months.
Comments: This ornamental grass is cultivated for garden effect and to be cut and dried for winter bouquets. The seeds are listed by Park and Thompson & Morgan (*see also* Appendix).

BABY-BLUE-EYES: *See Nemophila*

BABY'S-BREATH: *See Gypsophila*

BACHELOR'S-BUTTON: *See Centaurea*

BALSAM: *See Impatiens*

BARBERTON DAISY: *See Gerbera*

BEARDTONGUE: *See Penstemon*

BEAR'S-BREECH: *See Acanthus*

BEGONIA, WAX

When to plant: For indoors, plant in any season. For flowering-sized wax (*B. semperflorens*) hybrids to plant outdoors, sow seeds 12 weeks before frost-free weather is expected in the spring.
Temperature: 70° F.

Special treatment: Seeds need light to sprout; sow on surface, and do not cover with planting medium.
Days to sprout: 15.
Light for seedlings: Up to a half-day of sun, or start in a fluorescent-light garden (15–16 hours of light daily).
Maturity time: First blooms in about 12 weeks.
Comments: Begonia seeds are dust-sized, but surprisingly easy to grow. The number of varieties is almost unlimited. The Butterfly hybrid *B. semperflorens* varieties have individual flowers to 3 inches across—as showy as single-flowered, tuberous-rooted hybrids—source, Park (*see also* Appendix). Other outstanding wax begonia hybrids are listed by Butcher, Suttons, and Thompson & Morgan (*see also* Appendix).

BELLFLOWER: *See Campanula*

BLACK-EYED SUSAN: *See Rudbeckia*

BLACK-EYED-SUSAN VINE: *See Thunbergia*

BLANKETFLOWER: *See Gaillardia*

BLAZING-STAR: *See Mentzelia*

BLUE DAISY: *See Felicia*

BLUE LACEFLOWER: *See Trachymene*

BLUE MARGUERITE: *See Felicia*

BORAGE: *See Borago*

BORAGO (borage)

When to plant: Indoors, 6 to 8 weeks before planting-out time in the spring; outdoors, as soon as the soil is warm and there is no danger of frost.

Temperature: 70° F.

Special treatment: Continual darkness, until seeds sprout, produces the most successful germination.

Days to sprout: 8.

Light for seedlings: Half-day or more of sun, or start in a fluorescent-light garden (15–16 hours of light daily).

Maturity time: First blooms in about 3 months.

Comments: Easy to grow from seeds, this old-fashioned herb bears star-shaped, lavender blue flowers over a long season. The seeds are available from most seed companies (*see also* Appendix).

BRACHYCOME (Swan River daisy)

When to plant: Indoors, 6 to 8 weeks before planting-out time in the spring; outdoors, as soon as the soil is warm and there is no danger of frost.

Temperature: 70° F.

Special treatment: None required.

Days to sprout: 15.

Light for seedlings: Half-day or more of sun, or start in a fluorescent-light garden (15–16 hours of light daily).

Maturity time: First blooms in about 3 months.

Comments: Easy to grow from seeds, *Brachycome* gives a long season of bloom outdoors in flower beds, a rock garden, or in containers. Seeds are listed by Burpee, Butcher, Park, and Thompson & Morgan (*see also* Appendix).

BRASSICA (ornamental or flowering cabbage; ornamental or flowering kale)

When to plant: Indoors, 4 to 6 weeks before planting-out time in the spring; outdoors, as soon as the soil can be worked in the spring.

Temperature: 60° F.

These **Browallia** *seedlings have been started in a fluorescent-light garden and transplanted to one-inch plastic cups; they are about eight weeks old.*

Special treatment: Seeds need light to sprout; sow on surface, and do not cover with planting medium.
Days to sprout: 10.
Light for seedlings: Half-day or more of sun, or start in a fluorescent-light garden (15–16 hours of light daily).
Maturity time: Plants turn into showy rosettes of variegated leaves at the onset of cooler weather in autumn.
Comments: Easy to grow from seeds, *Brassica* may be grown in the ground outdoors or in containers. When fully colored, the plants may be dug up, potted, watered well, and brought indoors—discard when they are no longer attractive. The seeds are listed by Nichols and Park (*see also* Appendix).

BRIZA (quaking grass)

When to plant: Outdoors in spring as soon as the soil can be worked.
Temperature: 50°–60° F.
Special treatment: None required.
Days to sprout: 10.
Light for seedlings: Half-day or more of sun.
Maturity time: Graceful spikelets of seeds form in 3 to 5 months.
Comments: This ornamental grass is cultivated for garden effect and to be cut and dried for winter bouquets. The seeds are listed by Burpee, Park, and Thompson & Morgan (*see also* Appendix).

BROWALLIA (bush violet)

When to plant: For indoors, plant in any season. For flowering-sized seedlings to plant outdoors, sow seeds indoors 10 to 12 weeks before frost-free weather.
Temperature: 70° F.
Special treatment: Seeds need light to germinate; sow on the

surface of the planting medium.
Days to sprout: 15.
Light for seedlings: Up to a half-day of sun, or start in a fluorescent-light garden (15–16 hours of light daily).
Maturity time: 12 weeks for the first blue or white flowers.
Comments: As a winter-flowering house plant, *Browallia* requires abundant sunlight, temperatures of 55°–70° F., and fresh, moist air that circulates freely. Outdoors it makes a beautiful flowering plant that tolerates partial shade during warm weather; grow in the ground or in pots, tubs, window boxes, or hanging baskets. The seeds are listed by Burpee, Butcher, Park, Suttons, and Thompson & Morgan (*see also* Appendix).

BUGLOSS: *See Anchusa*

BURNING BUSH: *See Kochia*

BUSH VIOLET: *See Browallia*

BUSY LIZZY: *See Impatiens*

BUTTER DAISY: *See Verbesina*

BUTTERFLY-FLOWER: *See Schizanthus*

BUTTERFLY-PEA: *See Clitoria*

CABBAGE, ORNAMENTAL OR FLOWERING: *See Brassica*

CALCEOLARIA (pocketbook plant)

When to plant: Indoors in January for bloom the following summer.

Special treatment: Seeds need light to germinate; sow on the surface of the planting medium.
Temperature: 60°–70° F.
Days to sprout: 15.
Light for seedlings: Up to a half-day of sun, or start in a fluorescent-light garden (15–16 hours of light daily).
Maturity time: 6 months to first blooms.
Comments: The best *Calceolarias* to grow outdoors as summer-blooming annuals are hybrids of *C. rugosa*, especially Sunshine which is listed by Suttons (*see also* Appendix). *Calceolarias* of this type are also listed by Thompson & Morgan under the heading "Hardy and Half-hardy Annuals."

CALENDULA (pot-marigold)

When to plant: Indoors, 6 to 8 weeks before planting-out time in the spring; outdoors, as soon as the soil can be worked.
Temperature: 70° F.
Special treatment: Continual darkness, until seeds sprout, produces the most successful germination.
Days to sprout: 10.
Light for seedlings: Half-day or more of sun, or start in a fluorescent-light garden (15–16 hours of light daily).
Maturity time: First blooms in about 3 months.
Comments: Easy to grow from seeds, *Calendulas* bloom best in the cooler weather of late spring and early summer and again in early fall. For best results, pick spent flowers before seeds form. *Calendulas* are available from most seed companies (*see also* Appendix).

CALIFORNIA-POPPY: *See Eschscholzia*

CALLIOPSIS: *See Coreopsis*

CALLISTEPHUS (China aster; annual aster)

When to plant: Indoors, 6 to 8 weeks before planting-out time in the spring; outdoors, as soon as the soil is warm and there is no danger of frost.
Temperature: 70° F.
Special treatment: None required.
Days to sprout: 8.
Light for seedlings: A half-day or more of sun, or start in a fluorescent-light garden (15–16 hours of light daily).
Maturity time: First blooms in 3 to 4 months.
Comments: For best results, seedlings of China aster should never be allowed to dry out to the point of wilting the leaves; it is also important not to let the seedlings become rootbound before transplanting to the garden. China aster seeds are available from most seed companies (*see also* Appendix).

CAMPANULA (bellflower)

When to plant: Indoors, 8 to 12 weeks before planting-out time in the spring.
Temperature: 70°–75° F.
Special treatment: None required.
Days to sprout: 20.
Light for seedlings: Half-day or more of sun, or start in a fluorescent-light garden (15–16 hours of light daily).
Maturity time: First blooms in 4 to 6 months.
Comments: Most *Campanulas* are biennials or perennials; one annual of the Canterbury-bell type is listed by Park (*see also* Appendix).

CANDYTUFT: *See Iberis*

CAPE-FUCHSIA: *See Phygelius*

Callistephus *(China aster) seedling six weeks old and ready to plant in the garden.*

CAPE LEADWORT: *See Plumbago*

CAPE-MARIGOLD: *See Dimorphotheca*

CAPSICUM (Christmas pepper; ornamental pepper)

When to plant: Indoors, 6 to 8 weeks before planting-out time in the spring; outdoors, as soon as the soil is warm and there is no danger of frost.
Temperature: 70° F.
Special treatment: None required.
Days to sprout: 20.
Light for seedlings: Half-day or more of sun, or start in a fluorescent-light garden (15–16 hours of light daily).
Maturity time: First blooms in 3 months, followed by peppers—at first green, but maturing to brilliant red, orange, yellow, or purple.
Comments: These vary in shape from that of a cone to that of a chili pepper—but in miniature. The peppers are edible but very hot. Seeds of many unusual hybrids are listed by Butcher, Park, Suttons, and Thompson & Morgan (*see also* Appendix).

CASTOR-BEAN: *See Ricinus*

CATHARANTHUS (Madagascar periwinkle)

When to plant: Indoors, 8 to 10 weeks before planting-out time in the spring; outdoors, as soon as the soil is warm and there is no danger of frost.
Temperature: 70°–75° F.
Special treatment: Continual darkness, until seeds sprout, produces the most successful germination.
Days to sprout: 15.

Catharanthus, *usually called periwinkle or vinca, looks like this when about six weeks old; these seedlings might be transplanted to individual three-inch pots or directly to the garden.*

Light for seedlings: Half-day or more of sun, or start in a fluorescent-light garden (15–16 hours of light daily).

Maturity time: First blooms in 3 to 4 months.

Comments: Easy to grow, seeds of various hybrid Madagascar periwinkles are listed by most seed companies (*see also* Appendix) under the former name *Vinca rosea*.

CELOSIA (cockscomb)

When to plant: Indoors, 8 to 10 weeks before planting-out time in the spring; outdoors, as soon as the soil is warm and there is no danger of frost.

Temperature: 70°–75° F.

Special treatment: None required.

Days to sprout: 10.

Light for seedlings: Half-day or more of sun, or start in a fluorescent-light garden (15–16 hours of light daily).

Maturity time: First blooms in about 3 months.

Comments: Easy to grow, seeds of various hybrid *Celosias* (in both the crested and plume types) are listed in most catalogs (*see also* Appendix).

CELSIA

When to plant: Indoors, 8 to 12 weeks before planting-out time in the spring.

Temperature: 50°–60° F.

Special treatment: None required.

Days to sprout: 14–28.

Light for seedlings: Half-day or more of sun, or start in a fluorescent-light garden (15–16 hours of light daily).

Maturity time: First blooms in about 5 months.

Comments: This tender perennial is rarely seen in gardens but its purple-anthered, yellow flowers on long stems are very showy. *Celsia* seeds started in June or July will yield flowering

Celosia *seeds will grow to this size in about four weeks, at which time they can be transplanted outdoors provided the weather is warm and settled.*

plants for fall and winter in a cool-to-moderate home greenhouse. The seeds are listed by Butcher, Park, and Suttons (*see also* Appendix).

CENTAUREA (cornflower; bachelor's-button; dusty miller)

When to plant: If dusty miller, sow indoors 8 to 12 weeks before planting-out time in the spring. If cornflower or bachelor's-button (*Centaurea cyanus* hybrids), sow the seeds outdoors where they are to grow as soon as the soil is warm and there is no danger of frost.
Temperature: 65° F.
Special treatment: Continual darkness, until seeds sprout, produces the most successful germination.
Days to sprout: 10.
Light for seedlings: Half-day or more of sun; dusty-miller types may be started in a fluorescent-light garden (15-16 hours of light daily).
Maturity time: Dusty miller seedlings form attractive foliage plants within 3 months; allow 3 months for first blooms on cornflower or bachelor's-button.
Comments: Easy to grow, *Centaurea* seeds are listed in most catalogs (*see also* Appendix).

CENTRATHERUM (Manaos beauty)

When to plant: For indoors, plant in any season. For flowering-sized seedlings to plant outdoors, sow seeds 8 to 12 weeks before frost-free weather.
Temperature: 70° F.
Special treatment: None required.
Days to sprout: 10.
Light for seedlings: Half-day of sun, or grow in a fluorescent-light garden (15–16 hours of light daily).

Centaurea *(bachelor's-button)* seedlings reach this size in six weeks and can then be transplanted to the garden.

Maturity time: 3 months before the first lavender blue tassel flowers appear.

Comments: The leaves of this easily grown plant give off a delicious scent of pineapple when rubbed or brushed against. The name *Centratherum* is of unknown origin and cannot be verified in standard references such as *Hortus Third*. Seeds are listed by Park (*see also* Appendix).

CHAIN-OF-LOVE: *See Antigonon*

CHERRY-PIE: *See Heliotropium*

CHINA ASTER: *See Callistephus*

CHRISTMAS PEPPER: *See Capsicum*

CHRYSANTHEMUM (annual)

When to plant: Indoors, 6 to 8 weeks before planting-out time in the spring; outdoors, as soon as the soil is warm and there is no danger of frost.

Temperature: 65°–70° F.

Special treatment: None required.

Days to sprout: 8.

Light for seedlings: Half-day or more of sun, or start in a fluorescent-light garden (15–16 hours of light daily).

Maturity time: First blooms in about 3 months.

Comments: Easy from seeds, annual *Chrysanthemums* bloom best in the cooler weather of late spring and early summer and again in early fall. For best results, pick spent flowers before seeds form. Seeds of annual *Chrysanthemums* are listed in most catalogs (*see also* Appendix).

CIGARFLOWER: *See Cuphea*

CINERARIA: *See Senecio*

CLADANTHUS (Palm Springs daisy)

When to plant: Indoors, 8 to 10 weeks before planting-out time in the spring; outdoors, as soon as the soil is warm and there is no danger of frost.
Temperature: 70° F.
Special treatment: None required.
Days to sprout: 30.
Light for seedlings: Half-day or more of sun, or start in a fluorescent-light garden (15–16 hours of light daily).
Maturity time: First blooms in about 4 months.
Comments: Easy to grow from seeds, *Cladanthus* forms low mounds of finely cut leaves and bears a profusion of 2-inch golden yellow flowers all summer. *Cladanthus* is listed by Park and Thompson & Morgan (*see also* Appendix).

CLARKIA (farewell-to-spring; satin flower)

When to plant: Outdoors, where they are to grow and mature, as soon as the soil is warm and there is no danger of frost.
Temperature: 70° F.
Special treatment: None required.
Days to sprout: 5.
Light for seedlings: Half-day or more of sun.
Maturity time: First blooms in about 2 months.
Comments: This beautiful annual flower needs sun and sandy, well-drained soil. It performs best in the cooler days of spring and early summer and again in early fall. The seeds are listed by Burpee, Butchers, Park, Suttons, and Thompson & Morgan (*see also* Appendix).

CLEOME (spiderflower)

When to plant: Indoors, 8 to 10 weeks before planting-out time in the spring; outdoors, as soon as the soil can be worked.
Temperature: 65°–70° F.
Special treatment: Seeds need light to germinate; sow on the surface of the planting medium.
Days to sprout: 14–28.
Light for seedlings: Half-day or more of sun, or start in a fluorescent-light garden (15–16 hours of light daily).
Maturity time: Allow 3 to 4 months for first blooms.
Comments: Hybrid forms of this North American native make showy, drought-resistant garden flowers. The seeds are listed by Burpee, Butcher, Park, Suttons, and Thompson & Morgan (*see also* Appendix).

CLITORIA (butterfly-pea)

When to plant: Anytime.
Temperature: 70°–75° F.
Special treatment: Soak seeds in water at room temperature 24–48 hours before sowing.
Days to sprout: 15.
Light for seedlings: Half-day or more of sun, or start in a fluorescent-light garden (15–16 hours of light daily).
Maturity time: First blooms in 6 to 12 months.
Comments: This beautiful flowering vine behaves as a perennial in tropical and subtropical climates; elsewhere it may be started indoors about 12 weeks before warm, frost-free weather is expected outdoors. *Clitoria* also makes a choice flowering vine to grow in a moderate-to-warm greenhouse. The seeds are listed by Park and Thompson & Morgan (*see also* Appendix).

Cleome *seedlings reach this size in about eight weeks and can then be transplanted outdoors.*

COBAEA (cup-and-saucer vine)

When to plant: Indoors, 6 to 8 weeks before planting-out time in the spring; outdoors, as soon as the soil is warm and there is no danger of frost.
Temperature: 70° F.
Special treatment: None required.
Days to sprout: 15.
Light for seedlings: Half-day or more of sun, or start in a fluorescent-light garden (15–16 hours of light daily).
Maturity time: First blooms in about 3 months.
Comments: Easy to grow from seeds, *Cobaea* is listed, along with other annual vines, in most seed catalogs (*see also* Appendix).

COCKSCOMB: *See Celosia*

COLEUS

When to plant: For indoors, plant in any season. For showy foliage to plant outdoors in partial shade, sow seeds 8 to 12 weeks before frost-free weather.
Temperature: 65° F.
Special treatment: Seeds need light to germinate; sow on the surface of the planting medium.
Days to sprout: 10.
Light for seedlings: Half-day of sun or grow in a fluorescent-light garden (15–16 hours of light daily).
Maturity time: Seedlings show colorful foliage variegation within a few weeks' time; specimens may be grown in 6 months.
Comments: *Coleus* is an easy and rewarding plant to grow from seed. Many different hybrid strains are readily available (*see also* Appendix).

CONSOLIDA (larkspur)

When to plant: Sow in late fall or earliest spring in the area where the plants are to grow and bloom.
Temperature: 55° F.
Special treatment: Continual darkness, until seeds sprout, produces the most successful germination.
Days to sprout: 20.
Light for seedlings: Half-day or more of sun.
Maturity time: Allow 2 to 3 months from first sprouting to first blooms.
Comments: Larkspur is easy to grow from seeds but difficult to transplant. Hybrid larkspur seeds are listed in most catalogs (*see also* Appendix).

CORALVINE: *See Antigonon*

COREOPSIS (calliopsis)

When to plant: Indoors, 8 to 10 weeks before planting-out time in the spring; outdoors, as soon as the soil is warm and there is no danger of frost.
Temperature: 70° F.
Special treatment: None required.
Days to sprout: 8.
Light for seedlings: Half-day or more of sun, or start in a fluorescent-light garden (15–16 hours of light daily).
Maturity time: First blooms in about 3 months.
Comments: Easy to grow from seeds, this annual form of *Coreopsis* makes an outstanding plant for flowers all summer; it tolerates drought. For best results, pick spent flowers before seeds form. Calliopsis seeds are listed by Burpee, Butcher, Park, and Thompson & Morgan (*see also* Appendix).

CORNFLOWER: *See Centaurea*

COSMOS

When to plant: Outdoors, where they are to grow and bloom, as soon as the soil is warm and there is no danger of frost.
Temperature: 65°–70° F.
Special treatment: None required.
Days to sprout: 5.
Light for seedlings: Half-day or more of sun.
Maturity time: Allow 2 to 3 months for first blooms to appear.
Comments: *Cosmos* is easy to grow from seeds. For best results, pick spent blooms before seeds form. Seeds are listed in most catalogs (*see also* Appendix).

CREEPING ZINNIA: *See Sanvitalia*

CREPIS

When to plant: Outdoors, where they are to grow and bloom, as soon as the soil is warm and there is no danger of frost.
Temperature: 65°–70° F.
Special treatment: None required.
Days to sprout: 5.
Light for seedlings: Half-day or more of sun.
Maturity time: Allow 2 to 3 months for first blooms to appear.
Comments: *Crepis* is easy to grow. For best results, pick spent blooms before seeds form. *Crepis* bears red and white tassel flowers and is listed by Park and Thompson & Morgan (*see also* Appendix).

CUCURBIT (ornamental gourd)

When to plant: Indoors, 6 to 8 weeks before planting-out time in the spring; outdoors, as soon as the soil is warm and there is no danger of frost.
Temperature: 70°–75° F.

Cosmos Sunset (1966)
ORANGE SCARLET (all American)

Cosmos seedlings have finely cut leaves; these are six weeks old and ready for the garden.

Special treatment: None required.
Days to sprout: 8.
Light for seedlings: Half-day or more of sun, or start in a fluorescent-light garden (15–16 hours of light daily).
Maturity time: About 5 months.
Comments: Although easy to grow from seeds, ornamental gourds are difficult to transplant. If you start plants indoors, sow seeds in individual peat pots so that the seedlings can be transplanted to the garden with a minimum of root disturbance. Gourds are listed in most seed catalogs (*see also* Appendix).

CUPFLOWER: *See Nierembergia*

CUP-AND-SAUCER VINE: *See Cobaea*

CUPHEA (cigarflower; firecracker plant)

When to plant: Indoors, 8 to 10 weeks before planting-out time in the spring; outdoors, as soon as the soil is warm and there is no danger of frost.
Temperature: 70°–75° F.
Special treatment: Seeds need light to sprout; sow on surface, and do not cover with planting medium.
Days to sprout: 8.
Light for seedlings: Half-day or more of sun, or start in a fluorescent-light garden (15–16 hours of light daily).
Maturity time: First blooms in about 3 months.
Comments: Easy to grow from seeds, *Cuphea* is an excellent annual to grow as a bedding plant or in pots, tubs, window boxes, or hanging baskets outdoors. The seeds are listed by Butcher, Park, Suttons, and Thompson & Morgan (*see also* Appendix).

DAHLBERG DAISY: *See Dyssodia*

DATURA (angel's-trumpet)

When to plant: Indoors, 8 to 10 weeks before planting-out time in the spring; outdoors, as soon as the soil is warm and there is no danger of frost.
Temperature: 70° F.
Special treatment: None required.
Days to sprout: 15.
Light for seedlings: Half-day or more of sun, or start in a fluorescent-light garden.
Maturity time: First blooms in about 3 months.
Comments: Easy to grow from seeds, *Datura* is actually a tender perennial, the same as *Impatiens* and common garden geranium. Therefore, if it is planted in a large pot or tub, the plant can be indoors over winter and brought to bloom much earlier the following year. The seeds are listed by Butcher, Park, and Thompson & Morgan (*see also* Appendix).

DIANTHUS (pink; sweet William)

When to plant: Indoors, 8 to 10 weeks before planting-out time in the spring; outdoors, as soon as the soil is warm and there is no danger of frost.
Temperature: 70° F.
Special treatment: None required.
Days to sprout: 5.
Light for seedlings: Half-day or more of sun, or start in a fluorescent-light garden (15–16 hours of light daily).
Maturity time: First blooms in about 3 months.
Comments: *Dianthus* is easy to grow from seeds. Annual, biennial, and perennial hybrid varieties are listed in most seed catalogs. If the seeds of any are started early indoors, flowering may occur during the first summer (*see also* Appendix).

DIDISCUS: *See Trachymene*

Dianthus *seedlings of all types—annual, biennial, and perennial— closely resemble these which are four weeks old and ready to be transplanted to community flats or individual pots.*

Annual Dianthus *like these grow from seeds into full bloom in about three months.* Courtesy All-America Selections.

DIMORPHOTHECA (Cape-marigold)

When to plant: Indoors, 8 to 10 weeks before planting-out time in the spring; outdoors, as soon as the soil is warm and there is no danger of frost.
Temperature: 70° F.
Special treatment: None required.
Days to sprout: 10.
Light for seedlings: Half-day or more of sun, or start in a fluorescent-light garden (15–16 hours of light daily).
Maturity time: First blooms in about 3 months.
Comments: This low-growing plant literally covers itself with 2-inch (or larger) daisy flowers in many colors, some vivid, others pastel or glistening white. It tolerates considerable drought. For best results, pick spent flowers before seeds form. *Dimorphotheca* is also excellent for winter and spring bloom in a cool-to-moderate, sunny, airy home greenhouse. The seeds are listed by Burpee, Butcher, Park, Suttons, and Thompson & Morgan (*see also* Appendix).

DOROTHEANTHUS (iceplant)

When to plant: Indoors, 8 to 12 weeks before planting-out time in the spring; outdoors, as soon as the soil can be worked and there is little danger of frost.
Temperature: 60°–65° F.
Special treatment: Continual darkness, until seeds sprout, produces the most successful germination.
Days to sprout: 15.
Light for seedlings: Half-day or more of sun, or start in a fluorescent-light garden (15–16 hours of light daily).
Maturity time: First blooms in about 4 months.
Comments: Easy to grow from seeds, iceplant is one of the best ground covers possible for a sunny site with well-drained soil; it is highly tolerant of drought once the roots are established. You will find iceplant listed in the catalogs of Butcher,

Dimorphotheca *seedlings reach this size in about five weeks and are ready to transplant outdoors.*

Park, Suttons, and Thompson & Morgan (*see also* Appendix) under its former Latin name of *Mesembryanthemum*.

DUSTY MILLER: *See Centaurea*

DUSTY MILLER: *See Senecio*

DYSSODIA (Dahlberg daisy)

When to plant: Indoors, 6 to 8 weeks before planting-out time in the spring; outdoors, as soon as the soil can be worked and there is little danger of frost.
Temperature: 65°–70° F.
Special treatment: None required.
Days to sprout: 15.
Light for seedlings: Half-day or more of sun, or start in a fluorescent-light garden (15–16 hours of light daily).
Maturity time: First blooms in about 3 months.
Comments: This plant forms a mound of finely cut, bright green foliage less than 12 inches tall that covers itself over a long season with half-inch, bright yellow daisy flowers. It is excellent for container gardens outdoors in warm weather, for bedding, or to grow in pots for winter flowers in a cool-to-moderate sunny greenhouse. You will find the seeds listed in Park (*see also* Appendix) under the Dahlberg daisy's former Latin name of *Thymophylla*.

ECHIUM (viper's bugloss)

When to plant: Indoors, 6 to 8 weeks before planting-out time in the spring; outdoors, as soon as the soil can be worked and there is little danger of frost.
Temperature: 65°–70° F.
Special treatment: Continual darkness, until seeds sprout,

produces the most successful germination.
Days to sprout: 8.
Light for seedlings: Half-day or more of sun, or start in a fluorescent-light garden (15–16 hours of light daily).
Maturity time: First blooms in about 3 months.
Comments: Easy to grow from seeds, *Echium* does especially well in a sunny, well-drained site and tolerates poor soil and drought. The seeds are listed by Butcher, Park, Suttons, and Thompson & Morgan (*see also* Appendix).

ELEVEN-O'CLOCK: *See Portulaca*

EMILIA (tassel-flower; Flora's-paintbrush)

When to plant: Indoors, 6 to 8 weeks before planting-out time in the spring; outdoors, as soon as the soil is warm and there is no danger of frost.
Temperature: 70° F.
Special treatment: Continual darkness, until seeds sprout, produces the most successful germination.
Days to sprout: 8.
Light for seedlings: Half-day or more of sun, or start in a fluorescent-light garden (15–16 hours of light daily).
Maturity time: First blooms in about 4 months.
Comments: Easy to grow from seeds, *Emilia* is beautiful in the garden, but especially treasured as a long-lasting cut flower. Once the roots are established, *Emilia* also tolerates considerable drought. The seeds are listed by Park and Thompson & Morgan (*see also* Appendix).

ESCHSCHOLZIA (California-poppy)

When to plant: Outdoors, where they are to grow and bloom, in late fall or earliest spring.
Temperature: 60° F.

Special treatment: None required.
Days to sprout: 10.
Light for seedlings: Half-day or more of sun.
Maturity time: First blooms in 2 to 3 months.
Comments: This North American native makes a glorious ground cover in a sunny, well-drained site. It does not do well in rich soil that is constantly moist and cannot tolerate transplanting. Hybrids in the usual California-poppy color, as well as many beautiful pastels, are listed by Burpee, Butcher, Park, Suttons, and Thompson & Morgan (*see also* Appendix).

EUPHORBIA (annual poinsettia; snow-on-the-mountain)

When to plant: Indoors, 6 to 8 weeks before planting-out time in the spring; outdoors, as soon as the soil can be worked and there is little danger of frost.
Temperature: 70° F.
Special treatment: None required.
Days to sprout: 15.
Light for seedlings: Half-day or more of sun, or start in a fluorescent-light garden (15–16 hours of light daily).
Maturity time: First blooms in about 3 months.
Comments: *Euphorbia heterophylla*, or annual poinsettia, is also known as Mexican fire plant and Japanese poinsettia; *E. marginata* is called snow-on-the-mountain. Once established, these relatives of the Christmas poinsettia grow lustily to 3 feet tall or more and equally broad. While the soil should be well-drained, *Euphorbia* will grow in either rich or poor soil. Once the roots are established, *Euphorbia* tolerates drought unusually well. The seeds of both are listed by Park and Thompson & Morgan (*see also* Appendix).

EVERLASTING: *See Helipterum*

EVERLASTING: *See Helichrysum*

EXACUM (German violet; Persian violet)

When to plant: Anytime.
Temperature: 70° F.
Special treatment: Seeds need light to germinate; sow on the surface of the planting medium.
Days to sprout: 15.
Light for seedlings: Up to a half-day of sun, or start and grow in a fluorescent-light garden (15–16 hours of light daily).
Maturity time: 4 to 6 months.
Comments: Star-shaped blue or white flowers are fragrant. Discard the plants at the end of the flowering season. *Exacum* is usually treated as a house plant, but if the seeds are started indoors 8 to 12 weeks before frost-free weather is expected outdoors, it makes an excellent summer-flowering plant for container plantings in a partly shaded site. The seeds are listed by Butcher, Park, Suttons, and Thompson & Morgan (*see also* Appendix).

FAREWELL-TO-SPRING: *See Clarkia*

FELICIA (blue daisy; blue marguerite; kingfisher-daisy)

When to plant: Indoors, 8 to 10 weeks before planting-out time in the spring; outdoors, as soon as the soil is warm and there is no danger of frost.
Temperature: 60°–65° F.
Special treatment: None required.
Days to sprout: 14–28.
Light for seedlings: Half-day or more of sun, or start in a fluorescent-light garden (15–16 hours of light daily).
Maturity time: First blooms in about 4 months.
Comments: Easy to grow from seeds, *Felicia* is one of the few low-growing annuals with blue flowers. It is excellent for

Exacum *seedlings reach this size in about eight weeks and any time afterward can be transplanted outdoors to a shaded, moist place.*

planting in the ground or for use in all kinds of containers outdoors in warm weather. *Felicia* is often carried over during the winter as a greenhouse plant. The seeds are listed by Butcher, Park, Suttons, and Thompson & Morgan (*see also* Appendix).

FEVERFEW: *See Matricaria*

FIRECRACKER PLANT: *See Cuphea*

FLAX: *See Linum*

FLORA'S-PAINTBRUSH: *See Emilia*

FLOSSFLOWER: *See Ageratum*

FLOWERING-MAPLE: *See Abutilon*

FLOWERING TOBACCO: *See Nicotiana*

FOUNTAIN PLANT: *See Amaranthus*

FOUR-O'CLOCK: *See Mirabilis*

GAILLARDIA (blanketflower)

When to plant: Indoors, 6 to 8 weeks before planting-out time in the spring; outdoors, as soon as the soil is warm and there is no danger of frost.
Temperature: 70°–75° F.
Special treatment: None required.
Days to sprout: 20.
Light for seedlings: Half-day or more of sun, or start in a fluorescent-light garden (15–16 hours of light daily).
Maturity time: First blooms in about 3 months.
Comments: Easy to grow from seeds, *Gaillardia* is an outstand-

ing plant for a sunny site with well-drained soil. Once the roots are established, it tolerates considerable drought. You will find seeds of annual *Gaillardias* listed in the catalogs of Burpee, Butcher, Park, and Suttons (*see also* Appendix).

GAZANIA (treasure flower)

When to plant: Indoors, 8 to 10 weeks before planting-out time in the spring; outdoors, as soon as the soil is warm and there is no danger of frost.
Temperature: 60° F.
Special treatment: Continual darkness, until seeds sprout, produces the most successful germination.
Days to sprout: 8.
Light for seedlings: Half-day or more of sun, or start in a fluorescent-light garden (15–16 hours of light daily).
Maturity time: First blooms in about 4 months.
Comments: This low-growing plant bears large daisy flowers, about 3 inches across, in many vivid colors. It is excellent for planting in a sunny, well-drained site or for setting in outdoor container gardens during warm weather. *Gazania* is often cultivated for winter flowers in a sunny greenhouse with nighttime temperatures between 55°–65° F. The seeds are listed by Butcher, Park, Suttons, and Thompson & Morgan (*see also* Appendix).

GERANIUM: *See Pelargonium*

GERBERA (Transvaal daisy; Barberton daisy; African daisy; veldt daisy)

When to plant: For indoors, plant in any season. For flowering-sized seedlings to plant outdoors, sow seeds 12 to 16 weeks before frost-free weather.
Temperature: 70° F.

Special treatment: Plant fresh seeds with the pointed end down, other end exposed. Seeds need light to germinate.
Days to sprout: 15.
Light for seedlings: Half-day or more of sun, or start in a fluorescent-light garden (15–16 hours of light daily).
Maturity time: First blooms in 4 to 5 months.
Comments: *Gerbera* is a tender perennial that may be treated as an annual by starting the seeds early indoors, as described above. The plants can be kept over winter in a frost-free frame outdoors, or indoors in an atmosphere that is cool and moist with as much direct sun as possible. There are hybrids available with either single or double flowers. The seeds are listed by Burpee, Butcher, Park, Suttons, and Thompson & Morgan (*see also* Appendix).

GERMAN VIOLET: *See Exacum*

GILIA: *See Ipomopsis*

GLOBE AMARANTH: *See Gomphrena*

GLOBE-DAISY: *See Globularia*

GLOBULARIA (globe-daisy)

When to plant: Outdoors, where they are to grow and bloom, in late fall or earliest spring.
Temperature: 60° F.
Special treatment: None required.
Days to sprout: 10.
Light for seedlings: Half-day or more of sun.
Maturity time: First blooms in 4 to 5 months.
Comments: This little blue-flowered daisy grows only about 6 inches tall. In milder climates and well-drained soil it may perform as a hardy perennial; elsewhere it can be counted on for flowers from late summer until frost, if the seeds are sown

as recommended above. *Globularia* seeds are listed in Park (*see also* Appendix).

GLORIOSA DAISY: *See Rudbeckia*

GOLDEN-CUP: *See Hunnemannia*

GOLDEN TASSEL: *See Microsperma*

GOMPHOCARPUS: *See Asclepias*

GOMPHRENA (globe amaranth)

When to plant: Indoors, 8 to 10 weeks before planting-out time in the spring; outdoors, as soon as the soil is warm and there is no danger of frost.
Temperature: 65°–70° F.
Special treatment: Continual darkness, until seeds sprout, produces the most successful germination.
Days to sprout: 15.
Light for seedlings: Half-day or more of sun, or start in a fluorescent-light garden (15–16 hours of light daily).
Maturity time: First blooms in about 3 months.
Comments: Easy to grow from seeds, this plant has showy, cerise pink or white flowers that may be used, fresh or dried, in bouquets. *Gomphrena* thrives in full sun and tolerates poor soil and drought. The seeds are listed by Burpee, Butcher, Park, Suttons, and Thompson & Morgan (*see also* Appendix).

GOURD, ORNAMENTAL: *See Cucurbit*

GYPSOPHILA (baby's-breath)

When to plant: Indoors, 6 to 8 weeks before planting-out time in the spring; outdoors, as soon as the soil is warm and there is no danger of frost.

Temperature: 70° F.

Special treatment: None required.

Days to sprout: 10.

Light for seedlings: Half-day or more of sun, or start in a fluorescent-light garden (15–16 hours of light daily).

Maturity time: First blooms in about 3 months.

Comments: Baby's-breath makes a beautiful effect in the garden, but is treasured mostly as a cut flower, either fresh or dried. Seeds of the annual form (there are also perennials) are listed by Burpee, Butcher, Park, Suttons, and Thompson & Morgan (*see also* Appendix).

HARDY GLOXINIA: *See Incarvillea*

HELIANTHEMUM (sunrose; rock rose)

When to plant: Indoors, 8 to 10 weeks before planting-out time in the spring; outdoors, as soon as the soil is warm and there is no danger of frost.

Temperature: 65°–70° F.

Special treatment: None required.

Days to sprout: 15.

Light for seedlings: Half-day or more of sun, or start in a fluorescent-light garden (15–16 hours of light daily).

Maturity time: First blooms in about 5 months.

Comments: In milder climates and well-drained soil, *Helianthemum* may behave as a perennial; elsewhere it can be cultivated successfully as an annual by starting the seeds early indoors, as outlined above. The seeds are listed by Burpee, Park, Suttons, and Thompson & Morgan (*see also* Appendix).

HELIANTHUS (sunflower)

When to plant: Outdoors, where they are to grow and bloom,

as soon as the soil is warm and there is no danger of frost.
Temperature: 70°–75° F.
Special treatment: None required.
Days to sprout: 5.
Light for seedlings: Half-day or more of sun.
Maturity time: First blooms in 2 to 3 months.
Comments: Easy to grow, ideally in full sun, sunflowers tolerate drought as well as considerable wetness in the soil. Seeds are listed in most catalogs (*see also* Appendix).

HELICHRYSUM (everlasting; immortelle; straw-flower)

When to plant: Indoors, 8 to 10 weeks before planting-out time in the spring; outdoors, as soon as the soil is warm and there is no danger of frost.
Temperature: 70° F.
Special treatment: Seeds need light to sprout; sow on surface, and do not cover with planting medium.
Days to sprout: 5.
Light for seedlings: Half-day or more of sun, or start in a fluorescent-light garden (15–16 hours of light daily).
Maturity time: First blooms in about 3 months.
Comments: *Helichrysum* is easy to grow from seeds. Of all the straw-flowers or everlastings, it is the most popular for garden color in the summer and for dried bouquets that look fresh all winter. The seeds are listed in most catalogs (*see also* Appendix).

HELIOTROPE: *See Heliotropium*

HELIOTROPIUM (heliotrope; cherry-pie)

When to plant: Indoors, 8 to 10 weeks before planting-out time in the spring.

Temperature: 70° F.
Special treatment: None required.
Days to sprout: 25.
Light for seedlings: Half-day or more of sun, or start in a fluorescent-light garden (15–16 hours of light daily).
Maturity time: First blooms in 3 to 4 months.
Comments: Heliotrope is easy to grow from seeds, provided they are started indoors where an evenly moist growing medium can be assured. This favorite among old-fashioned scented flowers is actually a tender perennial that is usually treated as an annual; favorite seedlings can be kept over the winter in a window garden or greenhouse with an atmosphere that is sunny, moist, airy, and on the cool side (50°–60° F.). The seeds are listed by Burpee, Butcher, Park, Suttons, and Thompson & Morgan (*see also* Appendix).

HELIPTERUM (everlasting; straw-flower; rhodanthe)

When to plant: Indoors, 8 to 10 weeks before planting-out time in the spring; outdoors, as soon as the soil is warm and there is no danger of frost.
Temperature: 70° F.
Special treatment: None required.
Days to sprout: 15.
Light for seedlings: Half-day or more of sun, or start in a fluorescent-light garden (15–16 hours of light daily).
Maturity time: First blooms in about 3 months.
Comments: Although easy to grow for color in the garden, *Helipterum* is cultivated mostly for flowers to cut and dry for long-lasting winter bouquets. The seeds are listed in most catalogs, usually along with other everlastings (*see also* Appendix).

HUNNEMANNIA (Mexican tulip-poppy; golden-cup)

When to plant: Indoors, 8 to 12 weeks before planting-out

time in the spring; outdoors, as soon as the soil is warm and there is no danger of frost.

Temperature: 70°–75° F.

Special treatment: Seeds need light to sprout; sow on surface, and do not cover with planting medium.

Days to sprout: 15.

Light for seedlings: Half-day or more of sun, or start in a fluorescent-light garden (15–16 hours of light daily).

Maturity time: First blooms in about 5 months.

Comments: *Hunnemannia* is easy to grow; for best results, provide rich, moist, well-drained soil in a sunny site. *Hunnemannia* is classed as a tender perennial and may live for more than one season in mild climates. It is spectacular in the company of blue larkspur or *Delphinium*. The seeds are listed by Butcher and Park (*see also* Appendix).

IBERIS (candytuft)

When to plant: Indoors, 8 to 10 weeks before planting-out time in the spring; outdoors, as soon as the soil is warm and there is no danger of frost.

Temperature: 70° F.

Special treatment: None required.

Days to sprout: 20.

Light for seedlings: Half-day or more of sun, or start in a fluorescent-light garden (15–16 hours of light daily).

Maturity time: First blooms in about 3 months.

Comments: As candytuft is easy to grow from seeds, you will find both annual and perennial types in most seed catalogs, specifically those of Burpee, Butcher, Park, Suttons, and Thompson & Morgan (*see also* Appendix).

ICEPLANT: *See Dorotheanthus*

IMMORTELLE: *See Xeranthemum*

IMMORTELLE: *See Helichrysum*

IMPATIENS (balsam; busy Lizzy; patient Lucy; patience-plant; sultana)

When to plant: Indoors, 8 to 12 weeks before planting-out time in the spring; outdoors, as soon as the soil is warm and there is no danger of frost.

Temperature: 70° F.

Special treatment: The seeds of *Impatiens wallerana* hybrids (the *Impatiens* most commonly cultivated as a house plant and as an outdoor plant for shaded areas during warm weather) need light to sprout; sow on surface, and do not cover with planting medium. No special treatment is required for *Impatiens balsamina* hybrids, an annual usually called balsam.

Days to sprout: 15.

Light for seedlings: Up to a half-day of sun, or start in a fluorescent-light garden (15–16 hours of light daily).

Maturity time: First blooms in about 2 months.

Comments: The seeds of annual *Impatiens*, or balsam, are relatively large and extremely easy to grow, even when planted directly in the garden. Those of the tender perennial *Impatiens* are tiny, but not difficult to start if sown on the surface of the planting medium and kept evenly moist. You will find a large selection of hybrid *Impatiens* listed in the catalogs of Burpee, Butcher, Park, Suttons, and Thompson & Morgan (*see also* Appendix).

INCARVILLEA (hardy gloxinia)

When to plant: Indoors, 8 to 10 weeks before planting-out time in the spring; outdoors, as soon as the soil can be worked and there is little danger of frost.

Temperature: 60°–70° F.

Special treatment: None required.

Days to sprout: 25 or more.

Light for seedlings: Half-day or more of sun, or start in a fluorescent-light garden (15–16 hours of light daily).
Maturity time: First blooms in about 5 months.
Comments: While in no way related to the true gloxinia of florists, *Incarvillea* flowers do resemble those of the slipper-flowered gloxinias. In mild winter climates and in well-drained soil, *Incarvillea* behaves as a perennial. Elsewhere it can be treated as an annual, or held during the winter in a frost-free frame or set as a container plant in a cool greenhouse. The seeds are listed by Butcher, Park, Suttons, and Thompson & Morgan (*see also* Appendix).

IPOMOEA (morning-glory)

When to plant: Outdoors, where they are to grow and bloom, as soon as the soil is warm and there is no danger of frost.
Temperature: 70° F.
Special treatment: Soak seeds in water at room temperature for 24 to 48 hours before planting.
Days to sprout: 5.
Light for seedlings: Half-day or more of sun.
Maturity time: First blooms in about 2 months.
Comments: Although easy to grow from seeds, morning-glories are almost impossible to transplant. Seeds are listed in most catalogs (*see also* Appendix).

IPOMOPSIS (standing cypress)

When to plant: Outdoors, where they are to grow and bloom, as soon as the soil can be worked in the spring.
Temperature: 60°–70° F.
Special treatment: None required.
Days to sprout: 10.
Light for seedlings: Half-day or more of sun.
Maturity time: Spikes of scarlet flowers in about 4 months.
Comments: This plant is listed in the catalogs of Butcher and

Park (*see also* Appendix) under its former name of *Gilia*. It is easy to grow from seeds but difficult to transplant. The scarlet trumpet flowers appear on spikes that grow to 6 feet. This plant is sometimes listed as a biennial, and the flowers may not appear until the second season.

JACOBINIA: *See Justicia*

JASMINE TOBACCO: *See Nicotiana*

JEWELS-OF-OPAR: *See Talinum*

JOSEPH'S-COAT: *See Amaranthus*

JUSTICIA (jacobinia)

When to plant: Indoors, 8 to 12 weeks before planting-out time in the spring.
Temperature: 70°–75° F.
Special treatment: Seeds need light to sprout; sow on surface, and do not cover with planting medium.
Days to sprout: 30.
Light for seedlings: Half-day of sun, or start in a fluorescent-light garden (15–16 hours of light daily).
Maturity time: First blooms in about 5 months.
Comments: In frost-free climates, *Justicia* may be grown outdoors all year; elsewhere it should be treated as an indoor/outdoor container plant, very much the same as geraniums and wax begonias. You will find the seeds listed in Park as *Justicia*, in Thompson & Morgan as jacobinia (*see also* Appendix).

KALE, ORNAMENTAL OR FLOWERING: *See Brassica*

KINGFISHER-DAISY: *See Felicia*

KOCHIA (summer cypress; burning bush)

When to plant: Indoors, 6 to 8 weeks before planting-out time in the spring; outdoors, as soon as the soil is warm and there is no danger of frost.
Temperature: 70° F.
Special treatment: None required.
Days to sprout: 15.
Light for seedlings: Half-day or more of sun, or start in a fluorescent-light garden (15–16 hours of light daily).
Maturity time: Bushes become large enough for landscape effect in about 3 months.
Comments: Easy to grow from seeds, this annual is grown for foliage effect. Each bush grows 2 to 3 feet tall and about 12 inches across. In summer the finely cut leaves are pale green; in autumn the color changes to a coppery crimson. *Kochia* seeds are listed in most catalogs (*see also* Appendix).

LANTANA

When to plant: Anytime.
Temperature: 50°–65° F.
Special treatment: None required.
Days to sprout: 30–50 or more.
Light for seedlings: Half-day or more of sun, or start in a fluorescent-light garden (15–16 hours of light daily).
Maturity time: First flowers 6 to 8 months after sprouting.
Comments: *Lantana* seeds sprout erratically, some lying dormant in the ground for as much as a year. Plantings made in late winter or early spring are more likely to germinate within a reasonable time than those sown during other seasons. *Lantana* seeds are listed by Butcher, Park, and Thompson & Morgan (*see also* Appendix).

LARKSPUR: *See Consolida*

LATHYRUS (sweet pea)

When to plant: Outdoors, where they are to grow and bloom, as soon as the soil can be worked in late winter or early spring.
Temperature: 55° F.
Special treatment: Germination is hastened by soaking the seeds in water at room temperature for 24 hours before planting.
Days to sprout: 15.
Light for seedlings: Half-day or more of sun.
Maturity time: First blooms in 2 to 3 months.
Comments: Sweet peas are virtually impossible to transplant. Choice hybrids are listed in most catalogs (*see also* Appendix).

LAVATERA (tree-mallow)

When to plant: Indoors, 8 to 10 weeks before planting-out time in the spring; outdoors, as soon as the soil is warm and there is no danger of frost.
Temperature: 70° F.
Special treatment: None required.
Days to sprout: 20.
Light for seedlings: Half-day or more of sun, or start in a fluorescent-light garden (15–16 hours of light daily).
Maturity time: First blooms in about 3 months.
Comments: Easy to grow from seeds, this annual grows 2 to 3 feet tall and produces spectacular, hibiscuslike flowers. *Lavatera* is listed by Butcher, Park, Suttons, and Thompson & Morgan (*see also* Appendix).

LAYIA (tidy-tips)

When to plant: Indoors, 6 to 8 weeks before planting-out time in the spring; outdoors, as soon as the soil is warm and there is no danger of frost.
Temperature: 70° F.

Special treatment: None required.
Days to sprout: 8.
Light for seedlings: Half-day or more of sun, or start in a fluorescent-light garden (15–16 hours of light daily).
Maturity time: First blooms in about 3 months.
Comments: Easy to grow from seeds, *Layia* makes a long-lasting cut flower. The large golden yellow flowers have petals tipped neatly in white. The seeds are listed by Butcher, Park, and Thompson & Morgan (*see also* Appendix).

LAZY DAISY: *See Aphanostephus*

LIMONIUM (statice)

When to plant: Indoors, 6 to 8 weeks before planting-out time in the spring; outdoors, as soon as the soil is warm and there is no danger of frost.
Temperature: 70° F.
Special treatment: Continual darkness, until seeds sprout, produces the most successful germination.
Days to sprout: 15.
Light for seedlings: Half-day or more of sun, or start in a fluorescent-light garden (15–16 hours of light daily).
Maturity time: First blooms in about 3 months.
Comments: This annual provides flowers that are showy in the garden and easily air-dried to make long-lasting winter bouquets. You will find *Limonium* listed in most catalogs as statice (*see also* Appendix).

LINARIA (toadflax)

When to plant: Indoors, 6 to 8 weeks before planting-out time in the spring; outdoors, as soon as the soil is workable and there is little danger of hard freezing.
Temperature: 65° F.

Limonium *(statice) seedlings look like this when they are about eight weeks old and ready to plant in the garden.*

Special treatment: None required.
Days to sprout: 10.
Light for seedlings: Half-day or more of sun, or start in a fluorescent-light garden (15–16 hours of light daily).
Maturity time: First blooms in about 3 months.
Comments: Easily grown from seeds, this low-growing annual bears snapdragonlike flowers in vivid colors and is excellent for edging, massing in borders, as a ground cover, or in containers. The seeds are listed by Burpee, Butcher, Park, Suttons, and Thompson & Morgan (*see also* Appendix).

LINUM (flax)

When to plant: Indoors, 8 to 10 weeks before planting-out time in the spring; outdoors, as soon as the soil is warm and there is no danger of frost.
Temperature: 70° F.
Special treatment: None required.
Days to sprout: 25.
Light for seedlings: Half-day or more of sun, or start in a fluorescent-light garden (15–16 hours of light daily).
Maturity time: First blooms in about 3 months.
Comments: Easy to grow from seeds, annual flax (in blue, red, and white with a carmine eye) is listed by Butcher, Park, Suttons, and Thompson & Morgan (*see also* Appendix).

LOBELIA

When to plant: Indoors, 8 to 12 weeks before planting-out time in the spring; outdoors, as soon as the soil is warm and there is no danger of frost.
Temperature: 70°–75° F.
Special treatment: Seeds need light to sprout; sow on surface, and do not cover with planting medium.
Days to sprout: 20.
Light for seedlings: Half-day or more of sun, or start in a

fluorescent-light garden (15–16 hours of light daily).
Maturity time: First blooms in about 3 months.
Comments: Annual *Lobelia* grows about 6 inches tall; use as an edger for flower beds or in pots, window boxes, or hanging baskets. Hybrid forms are listed in most catalogs (*see also* Appendix).

LOBULARIA (sweet alyssum)

When to plant: Indoors, 6 to 8 weeks before planting-out time in the spring; outdoors, as soon as the soil can be worked and there is little danger of hard freezing.
Temperature: 70° F.
Special treatment: None required.
Days to sprout: 5.
Light for seedlings: Half-day or more of sun, or start in a fluorescent-light garden (15–16 hours of light daily).
Maturity time: First blooms in 6 to 8 weeks.
Comments: Sweet alyssum is exceedingly easy to grow from seeds. For best results, clip spent flowerheads back before seeds form. Seeds are listed in most catalogs (*see also* Appendix).

LOVE-IN-A-MIST: *See Nigella*

LOVE-LIES-BLEEDING: *See Amaranthus*

LUPINE: *See Lupinus*

LUPINUS (lupine; Texas bluebonnet)

When to plant: Outdoors, where they are to grow and bloom, as soon as the soil can be worked in late winter or early spring.
Temperature: 55° F.

These Lobelia seedlings are about four weeks old and ready to be transplanted to community flats or individual pots for a few more weeks of growing before planting outdoors.

Special treatment: Soak the seeds in water at room temperature for 24 hours before planting.
Days to sprout: 20.
Light for seedlings: Half-day or more of sun.
Maturity time: First blooms in about 4 months.
Comments: Lupines are virtually impossible to transplant. Seeds of the annual types are listed by Burpee, Butcher, Park, Suttons, and Thompson & Morgan (*see also* Appendix).

MACHAERANTHERA (Tahoka-daisy)

When to plant: Outdoors, where they are to grow and bloom, as soon as the soil can be worked in early spring.
Temperature: 55° F.
Special treatment: None required.
Days to sprout: 30.
Light for seedlings: Half-day or more of sun.
Maturity time: The first gold-centered, lavender blue flowers appear in about 3 months and continue until frost.
Comments: Tahoka daisy is easy to grow and makes an excellent show in the garden as well as in bouquets of cut flowers. The seeds are listed by Park (*see also* Appendix).

MADAGASCAR PERIWINKLE: *See Catharanthus*

MANAOS BEAUTY: *See Centratherum*

MARIGOLD: *See Tagetes*

MARTYNIA: *See Proboscidea*

MARVEL-OF-PERU: *See Mirabilis*

MATILIJA POPPY: *See Romneya*

maranthus *is an annual grown for colorful foliage.*

Impatiens *is one of the best annuals for flowers in mostly shaded areas.*

Celosia *dries easily for long-lasting winter bouquets.*

Dimorphotheca *makes an excellent ground cover.*

Helianthus *grows from seed to eight feet tall in only a few weeks of warm weather.*

Calendula *flowers best in cool, sunny weather.*

Papavers *or Iceland poppies like these will grow nearly wild from seeds broadcast in a meadow.*

Portulaca *is an annual ground cover with vivid blooms.*

Tagetes *or large-flowered hybrid marigolds are fun to grow in large patio containers.*

Dustymiller *is a form of* senecio *grown for its silvery gray foliage.*

This border is composed entirely of annuals; the pink spires are those of salvia, *the yellow globes, marigolds.*

Alcea *or hollyhock is a biennial that is easy to grow from seeds.*

Liatris *is a hardy perennial that deserves to be much better known; excellent for cutting.*

Hemerocallis, *or daylilies are available now in many pastel colors, like this pink one.*

Corydalis *is a treasured wildflower for early spring blooms.*

Digitalis *or foxglove is one of the most beautiful spire-form flowers.*

*V*iolas *and pansies are favorite* biennials *for planting with spring* bulbs *like these daffodils.*

Delphiniums *are perhaps the garden's most spectacular spire-form flowers.*

Lilium *(the yellow lily shown) and* pinkish Astilbe *make good companions in a flower garden.*

Sedum spectabile *(foreground) is a hardy perennial,* coleus *(background), an annual; both are easy from seeds.*

Primula—*hardy primroses—like these give vivid bloom the second season, in early spring.*

Tuberous begonia *hybrids like this one require about six months from seed to first blooms.*

Monarda *flowers in August and makes a good bee plant.*

Dahlias *are probably the easiest of all bulb flowers to grow from seeds; some are even treated as annuals.*

MATRICARIA (feverfew)

When to plant: Indoors, 6 to 8 weeks before planting-out time in the spring; outdoors, as soon as the soil can be worked and there is little danger of hard freezing.
Temperature: 70° F.
Special treatment: Seeds need light to sprout; sow on surface, and do not cover with planting medium.
Days to sprout: 15.
Light for seedlings: Half-day or more of sun, or start in a fluorescent-light garden (15–16 hours of light daily).
Maturity time: First blooms in about 3 months.
Comments: *Matricaria* is easy to grow from seeds and is outstanding as a garden flower and for cutting. It tends to live over winter in mild climates, especially if the soil is well-drained. The seeds are listed by Burpee, Butcher, Park, Suttons, and Thompson & Morgan (*see also* Appendix).

MATTHIOLA (stock)

When to plant: Indoors, 6 to 8 weeks before planting-out time in the spring; outdoors, as soon as the soil is warm and there is no danger of frost.
Temperature: 70° F.
Special treatment: None required.
Days to sprout: 10.
Light for seedlings: Half-day or more of sun, or start in a fluorescent-light garden (15–16 hours of light daily).
Maturity time: First blooms in about 3 months.
Comments: Easy to grow from seeds, some newer stock hybrids may bloom in 8 weeks from the date of germination. All types are outstanding cut flowers and most are sweetly scented. The seeds are listed by Burpee, Butcher, Park, Suttons, and Thompson & Morgan (*see also* Appendix).

Matthiola *(stock) seedlings reach this size in four weeks and are then ready for the garden.*

MENTZELIA (blazing-star)

When to plant: Indoors, 6 to 8 weeks before planting-out time in the spring; outdoors, as soon as the soil can be worked.
Temperature: 60°–70° F.
Special treatment: None required.
Days to sprout: 10.
Light for seedlings: Half-day or more of sun, or start in a fluorescent-light garden (15–16 hours of light daily).
Maturity time: First blooms in about 3 months.
Comments: Blazing-star is easy to grow from seeds. You will find this plant listed in the catalogs of Butcher, Park, Suttons, and Thompson & Morgan under its former name of Bartonia (*see also* Appendix).

MEXICAN CREEPER: *See Antigonon*

MEXICAN SUNFLOWER: *See Tithonia*

MEXICAN TULIP-POPPY: *See Hunnemannia*

MICROSPERMA (golden tassel)

When to plant: Anytime.
Temperature: 70°–75° F.
Special treatment: None required.
Days to sprout: 15.
Light for seedlings: About a half-day of sun, or grow in a fluorescent-light garden (15–16 hours of light daily).
Maturity time: Yellow flowers in 5 to 6 months on small plants easily cultivated indoors all year or outdoors in warm weather, either in the ground or in containers.
Comments: This little plant deserves to be much better known; the seeds are listed by Park (*see also* Appendix). The name *Microsperma* is of unknown origin and cannot be verified in standard references such as *Hortus Third*.

MIGNONETTE: *See Reseda*

MIMULUS (monkey flower)

When to plant: Indoors, 8 to 12 weeks before planting-out time in the spring.
Temperature: 65°–70° F.
Special treatment: None required.
Days to sprout: 10.
Light for seedlings: Half-day or more of sun, or start in a fluorescent-light garden (15–16 hours of light daily).
Maturity time: First blooms in about 3 months.
Comments: *Mimulus* seeds are tiny; merely press them into the surface of a moist seed-starting medium. This annual requires a relatively cool climate and soil that is rich and evenly moist. *Mimulus* is listed by Butcher, Park, Suttons, and Thompson & Morgan (*see also* Appendix).

MIRABILIS (four-o'clock; marvel-of-Peru)

When to plant: Indoors, 6 to 8 weeks before planting-out time in the spring; outdoors, as soon as the soil is warm and there is no danger of frost.
Temperature: 70°–75° F.
Special treatment: None required.
Days to sprout: 5.
Light for seedlings: Half-day or more of sun, or start in a fluorescent-light garden (15–16 hours of light daily).
Maturity time: First blooms in about 3 months.
Comments: Easy to grow from seeds, this is an old-fashioned annual that is not much seen in present-day gardens, although it should be. You will find the seeds listed by Park and Thompson & Morgan (*see also* Appendix).

MONARCH-OF-THE-VELDT: *See Venidium*

MONKEY FLOWER: *See Mimulus*

MORNING-GLORY: *See Ipomoea*

NASTURTIUM: *See Tropaeolum*

NEMESIA

When to plant: Indoors, 8 to 10 weeks before planting-out time in the spring; outdoors, as soon as the soil is warm and there is no danger of frost.
Temperature: 65° F.
Special treatment: Continual darkness, until seeds sprout, produces the most successful germination.
Days to sprout: 5.
Light for seedlings: Half-day or more of sun, or start in a fluorescent-light garden (15–16 hours of light daily).
Maturity time: First blooms in about 3 months.
Comments: *Nemesia* can be an annual flower of extraordinary beauty; however, it will not long survive hot, dry weather. If you have a greenhouse that is cool, moist, airy, and sunny in the winter, *Nemesia* seeds started in August or September will give winter–spring bloom in pots. The seeds are listed by Burpee, Butcher, Park, Suttons, and Thompson & Morgan (*see also* Appendix).

NEMOPHILA (baby-blue-eyes)

When to plant: Outdoors, where they are to grow and bloom, in latest fall or earliest spring.
Temperature: 60° F.
Special treatment: None required.
Days to sprout: 10.
Light for seedlings: Half-day or more of sun.
Maturity time: First blooms in about 2 months from the time of sprouting.

Nemesia seedlings grow quickly to this size—within four weeks—and may be transplanted to pots or directly to the garden.

Comments: This is a hardy annual that is very easy to grow from seeds. For best results, it needs cool weather and moist soil. The cup-shaped flowers are blue with a white center. The seeds are listed by Burpee, Butcher, Park, Suttons, and Thompson & Morgan (*see also* Appendix).

NICANDRA (apple-of-Peru; shoofly plant)

When to plant: Indoors, 8 to 10 weeks before planting-out time in the spring.
Temperature: 70° F.
Special treatment: None required.
Days to sprout: 15.
Light for seedlings: Half-day or more of sun, or start in a fluorescent-light garden (15–16 hours of light daily).
Maturity time: First blooms in about 4 months (blossoms are pale blue and bell-shaped), followed by small apple-shaped fruits.
Comments: In climates that have a long, warm growing season, *Nicandra* can be planted outdoors in the spring after the soil is warm and there is no danger of frost. Elsewhere it needs an early start indoors in order to mature by late summer. The seeds are listed by Butcher, Park, Suttons, and Thompson & Morgan (*see also* Appendix).

NICOTIANA (flowering tobacco; jasmine tobacco)

When to plant: Indoors, 8 to 10 weeks before planting-out time in the spring; outdoors, as soon as the soil is warm and there is no danger of frost.
Temperature: 70° F.
Special treatment: Seeds need light to sprout; sow on surface, and do not cover with planting medium.
Days to sprout: 20.
Light for seedlings: Half-day or more of sun, or start in a fluorescent-light garden (15–16 hours of light daily).

Maturity time: First blooms in about 4 months.
Comments: *Nicotiana* is easy to grow from seeds and makes an outstanding flowering plant from early summer to frost. It is excellent for bedding purposes and also to grow in containers outdoors. The varieties with white flowers are the most likely to be fragrant. Seeds of improved hybrids are listed by Burpee, Butcher, Park, Suttons, and Thompson & Morgan (*see also* Appendix).

NIEREMBERGIA (cupflower)

When to plant: Indoors, 8 to 12 weeks before planting-out time in the spring; outdoors, as soon as the soil is warm and there is no danger of frost.
Temperature: 70°–75° F.
Special treatment: None required.
Days to sprout: 15.
Light for seedlings: Half-day of sun, or start in a fluorescent-light garden (15–16 hours of light daily).
Maturity time: First blooms in about 3 months.
Comments: *Nierembergia* is easy to grow from seeds, provided the growing medium is never allowed to dry out during the early stages of development. It is actually a tender perennial that is usually treated as an annual. *Nierembergia* grows about 6 inches tall and is available in varieties with either violet blue or white flowers; it is outstanding for use in container gardens outdoors, especially in window boxes and hanging baskets. The seeds are listed by Burpee, Butcher, Park, and Thompson & Morgan (*see also* Appendix).

NIGELLA (love-in-a-mist)

When to plant: Outdoors, where they are to grow and bloom, as soon as the soil can be worked in spring and there is little danger of hard freezing.
Temperature: 60° F.

NICOTIANA, (Ht. 10")
RED DEVIL (Full Sun)

Nicotiana *seedlings reach this size in about six weeks, provided the atmosphere is warm and moist. They can go directly to the garden at this size, or somewhat larger.*

Special treatment: None required.
Days to sprout: 8.
Light for seedlings: Half-day or more of sun.
Maturity time: First blooms in about 3 months.
Comments: *Nigella* is easy to grow but virtually impossible to transplant. The seeds are listed by Burpee, Butcher, Park, Suttons, and Thompson & Morgan (*see also* Appendix).

OXYPETALUM (southern star)

When to plant: Indoors, 8 to 10 weeks before planting-out time in the spring; outdoors, as soon as the soil is warm and there is no danger of frost.
Temperature: 70° F.
Special treatment: None required.
Days to sprout: 10.
Light for seedlings: Half-day or more of sun, or start in a fluorescent-light garden (15–16 hours of light daily).
Maturity time: First blooms in 3 to 4 months.
Comments: Easy to grow from seeds, this plant is among the few annuals with light blue flowers. It is available from Park (*see also* Appendix).

PAINTED-TONGUE: *See Salpiglossis*

PALM SPRINGS DAISY: *See Cladanthus*

PASSIFLORA (passion-flower)

When to plant: Indoors, 8 to 12 weeks before planting-out time in the spring.
Temperature: 70° F.
Special treatment: None required.
Days to sprout: 30.
Light for seedlings: Half-day or more of sun, or start in a

fluorescent-light garden (15–16 hours of light daily).
Maturity time: First blooms in about 6 months.
Comments: This vine is available in many different species and varieties; some are cold-hardy in the north, others are tropicals. They can be treated as annuals in northern gardens, or grown in containers that can be moved inside during cold weather. The seeds are listed by Butcher, Park, and Thompson & Morgan (*see also* Appendix).

PASSION-FLOWER: *See Passiflora*

PATIENCE-PLANT: *See Impatiens*

PATIENT LUCY: *See Impatiens*

PELARGONIUM (geranium)

When to plant: Indoors, 8 to 12 weeks before planting-out time in the spring.
Temperature: 70° F.
Special treatment: None required.
Days to sprout: 15 or more.
Light for seedlings: Half-day or more of sun, or start in a fluorescent-light garden (15–16 hours of light daily).
Maturity time: First blooms in 4 to 5 months.
Comments: Geraniums are easy to grow from seeds. There are many outstanding hybrid strains available in seed form from Burpee, Butcher, Park, Suttons, and Thompson & Morgan (*see also* Appendix).

PENSTEMON (beardtongue)

When to plant: Indoors, 8 to 12 weeks before planting-out time in the spring.
Temperature: 65° F.

Special treatment: Continual darkness, until seeds sprout, produces the most successful germination.
Days to sprout: 10.
Light for seedlings: Half-day or more of sun, or start in a fluorescent-light garden (15–16 hours of light daily).
Maturity time: First blooms in about 5 months.
Comments: *Penstemon* (also listed as *Pentstemon*) is variously listed in catalogs as a hardy annual or hardy perennial. If seeds are started early indoors, as outlined above, most of the large-flowered hybrid strains available will perform as annuals that bloom the first year. They are listed by Burpee, Butcher, Park, Suttons, and Thompson & Morgan (*see also* Appendix).

PEPPER, ORNAMENTAL: *See Capsicum*

PERILLA

When to plant: Indoors, 6 to 8 weeks before planting-out time in the spring; outdoors, as soon as the soil is warm and there is no danger of frost.
Temperature: 65° F.
Special treatment: Seeds need light to sprout; sow on surface, and do not cover with planting medium.
Days to sprout: 15.
Light for seedlings: Half-day or more of sun, or start in a fluorescent-light garden (15–16 hours of light daily).
Maturity time: Becomes an attractive foliage plant (dark maroon purple leaves) within 3 months.
Comments: Easy to grow from seeds, *Perilla* is listed by Butcher, Park, Suttons, and Thompson & Morgan (*see also* Appendix).

PERSIAN VIOLET: *See Exacum*

These Petunia *seedlings are five weeks old and ready to be transplanted to community flats or individual pots for another few weeks before being planted in the garden.*

PETUNIA

When to plant: Indoors, 8 to 12 weeks before planting-out time in the spring.
Temperature: 70° F.
Special treatment: Seeds need light to sprout; sow on surface, and do not cover with planting medium.
Days to sprout: 10.
Light for seedlings: Half-day or more of sun, or start in a fluorescent-light garden (15–16 hours of light daily).
Maturity time: First blooms in about 3 months.
Comments: *Petunia* is easy to grow from seeds, provided it is started indoors where light, temperature, and moisture can be controlled. Many varieties of hybrid *Petunia* are listed in most catalogs (*see also* Appendix).

PHLOX DRUMMONDII (annual phlox)

When to plant: Indoors, 8 to 10 weeks before planting-out time in the spring; outdoors, as soon as the ground can be worked and there is little danger of frost.
Temperature: 65° F.
Special treatment: Continual darkness, until seeds sprout, produces the most successful germination.
Days to sprout: 10.
Light for seedlings: Half-day or more of sun, or start in a fluorescent-light garden (15–16 hours of light daily).
Maturity time: First blooms in about 3 months.
Comments: Easy to grow from seeds, annual phlox is excellent as a ground cover outdoors in a sunny site with well-drained soil. It tolerates drought once the roots are established and is excellent for container gardens outdoors, especially window boxes and hanging baskets. Seeds of hybrid annual phlox are listed in most catalogs (*see also* Appendix).

PHYGELIUS (Cape-fuchsia)

When to plant: Indoors, 8 to 12 weeks before planting-out time in the spring.
Temperature: 70° F.
Special treatment: None required.
Days to sprout: 10.
Light for seedlings: Half-day or more of sun, or start in a fluorescent-light garden (15–16 hours of light daily).
Maturity time: First blooms in about 6 months.
Comments: This plant is said to be winter-hardy to about 10° F.; however, it can also be treated as an annual by starting the seeds early indoors, as outlined above. The red, tubular flowers appear on shrublike bushes to 3 feet tall. *Phygelius* seeds are listed by Park and Thompson & Morgan (*see also* Appendix).

PIMPERNEL: *See Anagallis*

PINCUSHION FLOWER: *See Scabiosa*

PINK: *See Dianthus*

PLUMBAGO (Cape leadwort)

When to plant: Indoors, 8 to 12 weeks before planting-out time in the spring.
Temperature: 75° F.
Special treatment: None required.
Days to sprout: 25.
Light for seedlings: Half-day or more of sun, or start in a fluorescent-light garden (15–16 hours of light daily).
Maturity times: The first of the pale blue flowers should appear in about 5 months.
Comments: *Plumbago* is a tender perennial often treated as an

annual. Besides its usefulness as a bedding plant outdoors in warm weather, *Plumbago* is also excellent for container gardens of all kinds. The seeds are listed by Butcher, Park, and Thompson & Morgan (*see also* Appendix).

POCKETBOOK PLANT: *See Calceolaria*

POINSETTIA, ANNUAL: *See Euphorbia*

POLYGONUM

When to plant: Indoors, 6 to 8 weeks before planting-out time in the spring; outdoors, as soon as the soil is warm and there is no danger of frost.
Temperature: 70° F.
Special treatment: None required.
Days to sprout: 20.
Light for seedlings: Half-day or more of sun, or start in a fluorescent-light garden (15–16 hours of light daily).
Maturity time: Effective ground cover in a sunny site within 3 months.
Comments: This easy-to-grow annual makes a carpet of pink bronze foliage. The seeds are listed by Park and Thompson & Morgan (*see also* Appendix).

POOR-MAN'S-ORCHID: *See Schizanthus*

PORTULACA (rose moss; sun plant; eleven-o'clock)

When to plant: Outdoors, where they are to grow and bloom, as soon as the soil is warm and there is no danger of frost.
Temperature: 70° F.
Special treatment: Continual darkness, until seeds sprout,

Portulaca *seedlings reach this size in eight weeks, at which time they can go into the garden.*

produces the most successful germination.

Days to sprout: 10.

Light for seedlings: Half-day or more of sun.

Maturity time: First blooms in about 2 to 3 months.

Comments: Easy to grow from seeds, *Portulaca* needs sandy, well-drained soil and full sun. Once established, it is very tolerant of drought and is an excellent ground cover. Hybrid forms are listed in most catalogs (*see also* Appendix).

POT-MARIGOLD: *See Calendula*

PROBOSCIDEA (unicorn flower)

When to plant: Indoors, 8 to 10 weeks before planting-out time in the spring; outdoors, as soon as the soil is warm and there is no danger of frost.

Temperature: 70° F.

Special treatment: None required.

Days to sprout: 20.

Light for seedlings: Half-day or more of sun, or start in a fluorescent-light garden (15–16 hours of light daily).

Maturity time: First of the pinkish apricot, slipper gloxinialike flowers in about 3 months.

Comments: Once the roots are established, this North American native is very tolerant of drought. It grows to 3 feet in height and width with large, velvety green, heart-shaped leaves. The flowers are followed by unusual seedpods that, when mature and dry, may be used in dried arrangements. The seeds are listed by Park (*see also* Appendix) under the name *Martynia*.

QUAKING GRASS: *See Briza*

REHMANNIA

When to plant: Indoors, 8 to 12 weeks before planting-out time in the spring.
Temperature: 70° F.
Special treatment: None required.
Days to sprout: 15.
Light for seedlings: Half-day of sun, or start in a fluorescent-light garden (15–16 hours of light daily).
Maturity time: First blooms in about 5 months.
Comments: This half-hardy perennial, often treated as an annual, grows to about 2 feet tall with slipper-shaped flowers that are reddish pink with orange spots in the throat. Outdoors it needs partial shade and rich, moist, well-drained soil. The seeds are listed by Butcher, Park, and Thompson & Morgan (*see also* Appendix).

RESEDA (mignonette)

When to plant: Outdoors, where they are to grow and bloom, as soon as the soil can be worked and there is little danger of hard freezing.
Temperature: 70° F.
Special treatment: None required.
Days to sprout: 5.
Light for seedlings: Half-day or more of sun.
Maturity time: The first fragrant flowers in about 2 months from the time of sprouting.
Comments: This old-fashioned annual is easy to grow but difficult to transplant. The seeds are listed by Burpee, Butcher, Park, Suttons, and Thompson & Morgan (*see also* Appendix).

RHODANTHE: *See Helipterum*

Rehmannia *is a beautiful annual that performs best if the seeds are started early indoors.* Courtesy George W. Park Seed Co., Inc.

Although common mignonette, or Reseda, *is an annual that resents transplanting, seeds started in individual compartments of a flat, or in pots, can be moved to the garden with a minimum of root disturbance.*

RICINUS (castor-bean)

When to plant: Outdoors, where they are to grow, as soon as the soil is warm and there is no danger of frost.
Temperature: 70° F.
Special treatment: None required.
Days to sprout: 15.
Light for seedlings: Half-day or more of sun.
Maturity time: Large foliage plants within 3 months, becoming small trees by early autumn.
Comments: Castor-bean seeds are poisonous. This fast-growing annual with large, tropical-appearing leaves, is very easy to grow. The seeds are listed by Butcher, Park, Suttons, and Thompson & Morgan (*see also* Appendix).

ROCK ROSE: *See Helianthemum*

ROMNEYA (Matilija poppy)

When to plant: Outdoors, where they are to grow and bloom, as soon as the soil is warm and there is no danger of frost.
Temperature: 70° F.
Special treatment: None required.
Days to sprout: 30.
Light for seedlings: Half-day or more of sun.
Maturity time: Fragrant, white, 4-inch flowers in 4 to 5 months.
Comments: This poppy may grow to 8 feet in ideal conditions; in mild climates it tends to be a perennial. The seeds are listed by Park and Thompson & Morgan (*see also* Appendix).

ROSE MOSS: *See Portulaca*

Rudbeckia *seedlings reach this size in four weeks, at which time they can go directly to the garden or into individual pots for continued growth.*

RUDBECKIA (black-eyed Susan; Gloriosa daisy)

When to plant: Indoors, 8 to 10 weeks before planting-out time in the spring; outdoors, as soon as the soil is warm and there is little or no danger of frost.
Temperature: 70°–75° F.
Special treatment: None required.
Days to sprout: 20.
Light for seedlings: Half-day or more of sun, or start in a fluorescent-light garden (15–16 hours of light daily).
Maturity time: First blooms in about 4 months.
Comments: The Gloriosa daisy is a tetraploid form of the black-eyed Susan that grows wild over much of North America; if the soil is well-drained, it may behave as a hardy perennial. The seeds are listed by Burpee, Butcher, Park, Suttons, and Thompson & Morgan (*see also* Appendix).

SAGE: *See Salvia*

SALPIGLOSSIS (painted-tongue)

When to plant: Indoors, 8 to 10 weeks before planting-out time in the spring; outdoors, as soon as the soil is warm and there is no danger of frost.
Temperature: 70°–75° F.
Special treatment: Continual darkness, until seeds sprout, produces the most successful germination.
Days to sprout: 15.
Light for seedlings: Half-day or more of sun, or start in a fluorescent-light garden (15–16 hours of light daily).
Maturity time: First blooms in about 3 months.
Comments: Easy to grow from seeds, this annual offers the potential of spectacular blooms provided the soil is fairly rich, moist, and well-drained. The seeds are listed by Burpee, Butcher, Park, Suttons, and Thompson & Morgan (*see also* Appendix).

SALVIA (sage)

When to plant: Indoors, 8 to 10 weeks before planting-out time in the spring; outdoors, as soon as the soil is warm and there is no danger of frost.
Temperature: 70° F.
Special treatment: Continual darkness, until seeds sprout, produces the most successful germination.
Days to sprout: 15.
Light for seedlings: Half-day or more of sun, or start in a fluorescent-light garden (15–16 hours of light daily).
Maturity time: First blooms in about 3 months.
Comments: The annual *Salvia* is easy to grow from seeds. Besides the more common scarlet (which is often difficult to use effectively with other flowers, owing to the vivid shade of red), consider the Blue Bedder variety which looks beautiful with almost all other flower colors. Hybrid *Salvia* seeds of various types are listed by Burpee, Butcher, Park, Suttons, and Thompson & Morgan (*see also* Appendix).

SANVITALIA (creeping zinnia)

When to plant: Outdoors, where they are to grow and bloom, as soon as the soil is warm and there is no danger of frost.
Temperature: 70° F.
Special treatment: Seeds need light to sprout; sow on surface, and barely cover with planting medium.
Days to sprout: 10.
Light for seedlings: Half-day or more of sun.
Maturity time: First blooms in about 2 months.
Comments: Easy to grow from seeds, this plant makes an excellent ground cover in a sunny site; once established, it tolerates drought. The yellow flowers are like small, semidouble zinnias. The seeds are listed by Butcher, Park, and Thompson & Morgan (*see also* Appendix).

SATIN FLOWER: *See Clarkia*

Blue-flowered annual Salvia *seedlings look like this at about eight weeks of age, at which time they are ready for the garden.*

Red-, white-, or pink-flowered annual Salvia seedlings look like this at about eight weeks of age, at which time they are ready for the garden.

SCABIOSA (pincushion flower)

When to plant: Indoors, 8 to 10 weeks before planting-out time in the spring; outdoors, as soon as the soil is warm and there is no danger of frost.
Temperature: 70°–75° F.
Special treatment: None required.
Days to sprout: 10.
Light for seedlings: Half-day or more of sun, or start in a fluorescent-light garden (15–16 hours of light daily).
Maturity time: First blooms in about 3 months.
Comments: This annual is easy to grow from seeds; varieties are available that grow from 1½–3 feet tall. This annual is outstanding for use as a cut flower. The seeds are listed by Burpee, Butcher, Park, Suttons, and Thompson & Morgan (*see also* Appendix).

SCHIZANTHUS (butterfly-flower; poor-man's-orchid)

When to plant: Indoors, 8 to 12 weeks before planting-out time in the spring.
Temperature: 60° F.
Special treatment: Continual darkness, until seeds sprout, produces the most successful germination.
Days to sprout: 20.
Light for seedlings: Half-day or more of sun, or start in a fluorescent-light garden (15–16 hours of light daily).
Maturity time: First blooms in about 3 months.
Comments: This annual can be spectacular in any climate where summer weather is not mostly hot and dry; seeds started elsewhere in late August or September can be grown for winter and spring flowers in a home greenhouse that is sunny, airy, moist, and cool during the winter. *Schizanthus* seeds are listed by Burpee, Butcher, Park, Suttons, and Thompson & Morgan (*see also* Appendix).

SENECIO (cineraria; dusty miller)

When to plant: Dusty miller types indoors, 8 to 12 weeks before planting-out time in the spring; outdoors, as soon as the soil is warm and there is no danger of frost. Cineraria types (like the blue-flowered one shown on the cover of this book) indoors or outdoors (in a protected place) anytime from May to August.
Temperature: 75° F.
Special treatment: Seeds need light to sprout; sow on surface, and do not cover with planting medium.
Days to sprout: 10.
Light for seedlings: Half-day or more of sun, or start in a fluorescent-light garden (15–16 hours of light daily).
Maturity time: Dusty miller types become effective gray foliage plants in about 3 months. Cineraria types, started from seeds in the summer, will give winter and spring bloom indoors in an environment that is cool, sunny, airy, and moist.
Comments: Easy to grow from seeds, both types of *Senecio* are listed by Burpee, Butcher, Park, Suttons, and Thompson & Morgan (*see also* Appendix).

SHOOFLY PLANT: *See Nicandra*

SNAPDRAGON: *See Antirrhinum*

SOUTHERN STAR: *See Oxypetalum*

SNOW-ON-THE-MOUNTAIN: *See Euphorbia*

SPIDERFLOWER: *See Cleome*

STANDING CYPRESS: *See Ipomopsis*

STAR-OF-TEXAS: *See Xanthisma*

STATICE: *See Limonium*

STOCK: *See Matthiola*

STRAW-FLOWER: *See Helichrysum*

STRAW-FLOWER: *See Helipterum*

SULTANA: *See Impatiens*

SUMMER CYPRESS: *See Kochia*

SUNFLOWER: *See Helianthus*

SUN PLANT: *See Portulaca*

SUNROSE: *See Helianthemum*

SWAN RIVER DAISY: *See Brachycome*

SWEET ALYSSUM: *See Lobularia*

SWEET PEA: *See Lathyrus*

SWEET WILLIAM: *See Dianthus*

TAGETES (marigold)

When to plant: Indoors, 6 to 8 weeks before planting-out time in the spring; outdoors, as soon as the soil is warm and there is no danger of frost.
Temperature: 70° F.
Special treatment: None required.
Days to sprout: 5.
Light for seedlings: Half-day or more of sun, or start in a fluorescent-light garden (15–16 hours of light daily).

Maturity time: First blooms in about 3 months.
Comments: One of the easiest of all flowers to grow from seeds, hybrid marigolds are listed in most catalogs (*see also* Appendix).

TAHOKA DAISY: *See Machaeranthera*

TALINUM (jewels-of-Opar)

When to plant: Indoors, 8 to 10 weeks before planting-out time in the spring; outdoors, as soon as the soil is warm and there is no danger of frost.
Temperature: 70° F.
Special treatment: None required.
Days to sprout: 15.
Light for seedlings: Half-day or more of sun, or start in a fluorescent-light garden (15–16 hours of light daily).
Maturity time: First blooms in about 3 months.
Comments: Easy to grow from seeds, this unusual annual has waxy, shiny green leaves and countless tiny pink flowers. Once established, *Talinum* tolerates considerable drought. The seeds are listed by Park (*see also* Appendix).

TASSEL-FLOWER: *See Emilia*

TEXAS BLUEBONNET: *See Lupinus*

THUNBERGIA (black-eyed-Susan vine)

When to plant: Indoors, 6 to 8 weeks before planting-out time in the spring; outdoors, as soon as the soil is warm and there is no danger of frost.
Temperature: 70°–75° F.

Special treatment: None required.
Days to sprout: 10.
Light for seedlings: Half-day or more of sun, or start in a fluorescent-light garden (15–16 hours of light daily).
Maturity time: First blooms in about 3 months.
Comments: This annual vine is easy to grow from seeds. Besides the more common orange yellow flowers with a dark eye, there is also a newer hybrid with pure white flowers. The seeds are listed by Burpee, Butcher, Park, Suttons, and Thompson & Morgan (*see also* Appendix).

TIDY-TIPS: *See Layia*

TITHONIA (Mexican sunflower)

When to plant: Outdoors, where they are to grow and bloom, as soon as the soil is warm and there is no danger of frost.
Temperature: 70° F.
Special treatment: Continual darkness, until seeds sprout, produces the most successful germination.
Days to sprout: 20.
Light for seedlings: Half-day or more of sun.
Maturity time: First blooms in about 2 months.
Comments: This tall-growing annual is easy to grow and makes a beautiful appearance in the garden; it is also excellent as a cut flower. The seeds are listed by Burpee, Park, Suttons, and Thompson & Morgan (*see also* Appendix).

TOADFLAX: *See Linaria*

TORENIA (wishbone flower)

When to plant: Indoors, 8 to 12 weeks before planting-out

Thunbergia (black-eyed-Susan vine) seedlings are best started in individual pots like this one. When about eight weeks old they can be moved to the garden with virtually no root disturbance.

time in the spring; outdoors, as soon as the soil is warm and there is no danger of frost.

Temperature: 70°–75° F.

Special treatment: None required.

Days to sprout: 15.

Light for seedlings: Half-day of sun, or start in a fluorescent-light garden (15–16 hours of light daily).

Maturity time: First blooms in about 3 months.

Comments: This annual is easy to grow from seeds and is excellent for flowers in a partly shaded area outdoors, either in the ground or in containers. The seeds are listed by Burpee, Park, Suttons, and Thompson & Morgan (*see also* Appendix).

TRACHYMENE (blue laceflower)

When to plant: Indoors, 8 to 10 weeks before planting-out time in the spring; outdoors, as soon as the soil is warm and there is no danger of frost.

Temperature: 65° F.

Special treatment: Continual darkness, until seeds sprout, produces the most successful germination.

Days to sprout: 15.

Light for seedlings: Half-day or more of sun, or start in a fluorescent-light garden (15–16 hours of light daily).

Maturity time: First blooms in about 3 months.

Comments: Easy from seeds, this annual is beautiful in the garden, but especially valued as a cut flower. The seeds are listed by Burpee, Butcher, Park, Suttons, and Thompson & Morgan (*see also* Appendix), usually under the former name of *Didiscus*.

TRANSVAAL DAISY: *See Gerbera*

TREASURE FLOWER: *See Gazania*

TREE MALLOW: *See Lavatera*

TROPAEOLUM (nasturtium)

When to plant: Outdoors, where they are to grow and bloom, as soon as the soil is warm and there is no danger of frost.
Temperature: 65° F.
Special treatment: Continual darkness, until seeds sprout, produces the most successful germination.
Days to sprout: 8.
Light for seedlings: Half-day or more of sun.
Maturity time: First blooms in about 2 months.
Comments: One of the easiest of all annual flowers to grow from seeds, nasturtium is difficult to transplant. Some varieties grow as compact bushes, others trail or climb (the canary-bird creeper, for example). Seeds are listed in most catalogs (*see also* Appendix).

UNICORN FLOWER: *See Proboscidea*

VELDT DAISY: *See Gerbera*

VENIDIUM (monarch-of-the-veldt)

When to plant: Indoors, 8 to 12 weeks before planting-out time in the spring; outdoors, as soon as the soil is warm and there is no danger of frost.
Temperature: 70° F.
Special treatment: Seeds need light to sprout; sow on surface, and do not cover with planting medium.
Days to sprout: 8.
Light for seedlings: Half-day or more of sun, or start in a fluorescent-light garden (15–16 hours of light daily).
Maturity time: First blooms in about 4 months.

Comments: This annual is excellent to grow in a hot, dry, sunny site; it bears beautifully colored daisy flowers, 5 inches across on 2½-foot stems. It is excellent for cutting. The seeds are listed by Butcher, Suttons, and Park (*see also* Appendix).

VERBENA

When to plant: Indoors, 8 to 10 weeks before planting-out time in the spring; outdoors, as soon as the soil is warm and there is no danger of frost.
Temperature: 65° F.
Special treatment: Continual darkness, until seeds sprout, produces the most successful germination.
Days to sprout: 20.
Light for seedlings: Half-day or more of sun, or start in a fluorescent-light garden (15–16 hours of light daily).
Maturity time: First blooms in about 3 months.
Comments: *Verbena* is one of the best annuals to use as a ground cover in a site that is sunny and well-drained; once established, it tolerates drought. Trailing varieties are beautiful when planted in window boxes or hanging baskets. The seeds are listed in most catalogs (*see also* Appendix).

VERBESINA (butter daisy)

When to plant: Outdoors, where they are to grow and bloom, as soon as the soil is warm and there is no danger of frost.
Temperature: 70° F.
Special treatment: None required.
Days to sprout: 8.
Light for seedlings: Half-day or more of sun.
Maturity time: First of the 2-inch, butter yellow daisy flowers in about 3 months.
Comments: This little-known annual is easy to grow from seeds and is excellent for garden color as well as for cutting to

Verbena *seedlings reach this size in about seven weeks; pot plants individually or move them directly to the garden.*

Xeranthemum seedlings started under fluorescent lights after about three weeks; in a few more days they will be large enough to space out in a community flat or to move to individual small pots.

use in bouquets. The plants grow to be about 3 feet tall. *Verbesina* seeds are listed by Park (*see also* Appendix).

VINCA: *See Catharanthus*

VIPER'S BUGLOSS: *See Echium*

WINGED EVERLASTING: *See Ammobium*

WISHBONE FLOWER: *See Torenia*

XANTHISMA (star-of-Texas)

When to plant: Indoors, 8 to 10 weeks before planting-out time in the spring; outdoors, as soon as the soil is warm and there is no danger of frost.
Temperature: 70° F.
Special treatment: None required.
Days to sprout: 25.
Light for seedlings: Half-day or more of sun, or start in a fluorescent-light garden (15–16 hours of light daily).
Maturity time: First blooms in about 3 months.
Comments: This annual produces large, pale yellow, single daisy flowers on 18-inch stems; they are excellent to use in bouquets. The seeds are listed by Park (*see also* Appendix).

XERANTHEMUM (immortelle)

When to plant: Outdoors, where they are to grow and bloom, as soon as the soil is warm and there is no danger of frost.
Temperature: 70° F.
Special treatment: None required.
Days to sprout: 10.
Light for seedlings: Half-day or more of sun.

Maturity time: First blooms in about 4 months.

Comments: This easy-to-grow annual is one of the everlastings, cultivated for use as a dried flower in winter bouquets. The seeds are listed by Burpee, Butcher, Park, Suttons, and Thompson & Morgan (*see also* Appendix).

YOUTH AND OLD AGE: *See Zinnia*

ZINNIA (youth and old age)

When to plant: Indoors, 6 to 8 weeks before planting-out time in the spring; outdoors, as soon as the soil is warm and there is no danger of frost.

Temperature: 70°–75° F.

Special treatment: None required.

Days to sprout: 5.

Light for seedlings: Half-day or more of sun, or start in a fluorescent-light garden (15–16 hours of light daily).

Maturity time: First blooms in about 3 months.

Comments: One of the easiest of all annuals to grow from seeds, *Zinnia* is available in a wide variety of plant sizes (from 6 to 36 inches tall), flower sizes, and blossom colors. The seeds are listed in most catalogs (*see also* Appendix).

Zinnia *seedlings look like this about 14 days after being planted.*

These Zinnia seedlings are about four weeks old and large enough to transplant to the garden or to individual three-inch pots.

Biennial Flowers to Grow from Seeds

It is almost as risky to pigeonhole plants as it is people; however, generally speaking, a biennial is a plant that spends all of its energy the first year growing leaves and a good root system, and the second year it blooms, sets seeds, and dies.

The list of biennial flowers reads like the roster for great-grandmother's cottage garden: Canterbury bells, foxglove, sweet William, English daisy, some forget-me-nots, hollyhock, sweet rocket, mullein, wallflower, Johnny-jump-up, and pansy. Great-grandmother either had more time or—more likely—was willing to work harder, so she grew the biennials. Their two-season cycle goes like this: sow seeds in a protected frame in late spring or summer; mulch well after the ground freezes in autumn; and as soon as the soil can be worked in early spring, transplant the biennials to where they will bloom.

Biennials behave differently in different gardens and different climates. Some may even return several seasons, revealing a more perennial nature. Some, sweet rocket in particular, self-seed; therefore, all you have to do is save selected volunteers and hoe up the others.

ALCEA (hollyhock)

When to plant: Outdoors, in a seedframe, in spring or summer; indoors, 6 to 8 weeks before planting-out time in the spring.

These Alcea *(hollyhock) seedlings were started outdoors in a protected seedframe in early summer. By late summer they can be planted directly in the garden where they are to mature and bloom the following year.*

Temperature: 70° F.
Special treatment: None required.
Days to sprout: 10.
Light for seedlings: Half-day or more of sun, or start in a fluorescent-light garden (15–16 hours of light daily).
Maturity time: First blooms may occur toward end of summer from seeds started early indoors, as described above; otherwise, in the second season.
Comments: Very easy to grow from seeds, hollyhocks have been so vastly improved in recent years that some of the newer hybrids may behave as annuals or perennials instead of biennials. The seeds are listed by Burpee, Butcher, Park, Suttons, and Thompson & Morgan (*see also* Appendix).

BELLIS (English daisy)

When to plant: Outdoors, in a seedframe, in spring or summer.
Temperature: 70° F.
Special treatment: Seeds need light to sprout; sow on surface, and do not cover with planting medium.
Days to sprout: 8.
Light for seedlings: Half-day of sun.
Maturity time: Blooms in winter or spring of second year.
Comments: Easy to grow, *Bellis* withers quickly in hot, dry weather. The seeds are listed by Burpee, Butcher, Park, Suttons, and Thompson & Morgan (*see also* Appendix).

CAMPANULA (Canterbury bells; cup and saucer)

When to plant: Outdoors, in a seedframe, in spring or summer.
Temperature: 70° F.
Special treatment: None required.
Days to sprout: 20.

Campanula (or Canterbury bell) is one of the most beautiful of all biennials. Flowers like this appear in spring and early summer approximately one year from the time seeds are sown.

Light for seedlings: Half-day or more of sun.
Maturity time: Blooms in early summer of second year.
Comments: Easy to grow from seeds, Canterbury bells usually have blue, white, or pink flowers. Well-grown plants in full bloom are among the garden's choicest biennials. Hybrid seeds are listed by Burpee, Butcher, Park, Suttons, and Thompson & Morgan (*see also* Appendix).

CANTERBURY BELLS: *See Campanula*

CHEIRANTHUS (wallflower)

When to plant: Outdoors, in a seedframe, in spring or summer.
Temperature: 70° F.
Special treatment: None required.
Days to sprout: 5.
Light for seedlings: Half-day or more of sun.
Maturity time: Blooms in spring and early summer of second year.
Comments: Easy to grow from seeds, wallflowers may not be hardy during the winter north of Philadelphia unless given the protection of a coldframe. Hybrid seeds are listed by Burpee, Butcher, Park, Suttons, and Thompson & Morgan (*see also* Appendix).

CHINESE FORGET-ME-NOT: *See Cynoglossum*

CUP AND SAUCER: *See Campanula*

CYNOGLOSSUM (Chinese forget-me-not)

When to plant: Outdoors, in a seedframe, in early spring while the soil is cool.

Temperature: 65° F.
Special treatment: Continual darkness, until seeds sprout, produces the most successful germination.
Days to sprout: 5.
Light for seedlings: Half-day of sun.
Maturity time: Blue, pink, or white forget-me-not flowers in spring and summer of second year.
Comments: Easy to grow from seeds, this biennial may behave as a perennial if summer temperatures are not too hot and the plants have adequate moisture and shade. The seeds are listed by Burpee, Butcher, Park, Suttons, and Thompson & Morgan (*see also* Appendix).

DAMES' ROCKET: *See Hesperis*

DAUCUS (Queen-Anne's-lace)

When to plant: Outdoors, where they are to grow and bloom, as soon as the soil is warm in the spring and there is no danger of frost; indoors, 6 to 8 weeks before planting-out time in the spring.
Temperature: 70° F.
Special treatment: Seeds need light to sprout; sow on surface, and do not cover with planting medium.
Days to sprout: 10.
Light for seedlings: Half-day or more of sun, or start in a fluorescent-light garden (15–16 hours of light daily).
Maturity time: Lacy, white flowers in summer of second year; some plants may bloom toward the end of the first season.
Comments: This wild flower is being invited into an increasing number of cultivated gardens because of its beauty, especially as cutting material for bouquets. The seeds are listed by Park (*see also* Appendix).

DIANTHUS (sweet William)

When to plant: Outdoors, in a seedframe, in spring or summer; indoors, 6 to 8 weeks before planting-out time in the spring.
Temperature: 70° F.
Special treatment: None required.
Days to sprout: 5.
Light for seedlings: Half-day or more of sun, or start in a fluorescent-light garden (15–16 hours of light daily).
Maturity time: Blooms in spring and summer of second year; seeds started early indoors, as described above, may yield some flowers toward the end of the first season.
Comments: Easy to grow from seeds, like most biennials, *Dianthus* is unusually beautiful. The flowers are lightly clove-scented and among the best for bouquets. The seeds are listed by Burpee, Butcher, Park, Suttons, and Thompson & Morgan (*see also* Appendix).

DIGITALIS (foxglove)

When to plant: Outdoors, in a seedframe, in spring or summer; indoors, 6 to 8 weeks before planting-out time in the spring.
Temperature: 70° F.
Special treatment: Seeds need light to sprout; sow on surface, and do not cover with planting medium.
Days to sprout: 20.
Light for seedlings: Half-day or more of sun, or start in a fluorescent-light garden (15–16 hours of light daily).
Maturity time: Blooms in spring and summer of second year. Some of the newer hybrids behave more as annuals and bloom in late summer of the first season, provided the seeds are given an early start indoors, as described above.
Comments: Easy to grow from seeds, foxglove is one of the garden's showiest spire-form flowers. The seeds are listed by

Burpee, Butcher, Park, Suttons, and Thompson & Morgan (*see also* Appendix).

ENGLISH DAISY: *See Bellis*

EVENING-PRIMROSE: *See Oenothera*

FORGET-ME-NOT: *See Myosotis*

FOXGLOVE: *See Digitalis*

HESPERIS (dames' rocket; sweet rocket)

When to plant: Outdoors, in a seedframe, in spring or summer; indoors, 6 to 8 weeks before planting-out time in the spring.

Temperature: 70° F.

Special treatment: Seeds need light to sprout; sow on surface, and do not cover with planting medium.

Days to sprout: 25.

Light for seedlings: Half-day or more of sun, or start in a fluorescent-light garden (15–16 hours of light daily).

Maturity time: Blooms in spring and summer of second year; seedlings given an early start indoors, as described above, may give some bloom toward the end of the first season.

Comments: Easy to grow from seeds, *Hesperis* is a biennial that readily perpetuates itself by self-sown seedlings. However, for the best garden effect, it is better to collect seeds from the choicest plants and start them. The seeds are listed by Park and Thompson & Morgan (*see also* Appendix).

HOLLYHOCK: *See Alcea*

HONESTY: *See Lunaria*

JOHNNY-JUMP-UP: *See Viola*

LUNARIA (honesty; money plant)

When to plant: Outdoors, in a seedframe, in spring or summer; indoors, 6 to 8 weeks before planting-out time in the spring.
Temperature: 65°–70° F.
Special treatment: None required.
Days to sprout: 10.
Light for seedlings: Half-day or more of sun.
Maturity time: Blooms in spring and summer of second year, followed by the seedpods which dry to become "money."
Comments: *Lunaria* is easy to grow. Seeds are listed by Burpee, Park, Suttons, and Thompson & Morgan (*see also* Appendix).

MONEY PLANT: *See Lunaria*

MULLEIN: *See Verbascum*

MYOSOTIS (forget-me-not)

When to plant: Outdoors, in a seedframe, in spring or summer; indoors, 6 to 8 weeks before planting-out time in the spring.
Temperature: 65°–70° F.
Special treatment: Continual darkness, until the seeds sprout, produces the most successful germination.
Days to sprout: 8.
Light for seedlings: Half-day of sun, or start in a fluorescent-light garden (15–16 hours of light daily).
Maturity time: Blooms in spring and summer of second year; seedlings started early indoors, as described above, may give

some bloom in late summer and fall of the first season.

Comments: Easy to grow from seeds, *Myosotis* is prized as a companion flower for spring bulbs such as tulips and hyacinths. The seeds are listed by Burpee, Butcher, Park, Suttons, and Thompson & Morgan (*see also* Appendix).

OENOTHERA (evening-primrose)

When to plant: Outdoors, in a seedframe, in spring or summer; indoors, 6 to 8 weeks before planting-out time in the spring.

Temperature: 70° F.

Special treatment: None required.

Days to sprout: 5.

Light for seedlings: Half-day or more of sun, or start in a fluorescent-light garden (15–16 hours of light daily).

Maturity time: Blooms in summer of second year; seedlings started early indoors, as described above, may give some bloom in late summer and fall of the first season.

Comments: *Oenothera* is easy to grow from seeds; once established, it tolerates drought. Evening-primroses perform variously as annuals, biennials, or perennials, depending on the environment as well as the genetic makeup of the seeds. *Oenothera* seeds are listed by Butcher, Park, Suttons, and Thompson & Morgan (*see also* Appendix).

PANSY: *See Viola*

QUEEN-ANNE'S-LACE: *See Daucus*

SWEET ROCKET: *See Hesperis*

VERBASCUM (mullein)

When to plant: Outdoors, in a seedframe, in spring or sum-

mer; indoors, 6 to 8 weeks before planting-out time in the spring.
Temperature: 70° F.
Special treatment: None required.
Days to sprout: 20.
Light for seedlings: Half-day or more of sun, or start in a fluorescent-light garden (15–16 hours of light daily).
Maturity time: Blooms in spring and summer of second year.
Comments: Easily grown from seeds, *Verbascum* is one of the garden's most beautiful spire-form flowers. Hybrid seeds are listed by Butcher, Park, Suttons, and Thompson & Morgan (*see also* Appendix).

VIOLA (pansy; Johnny-jump-up)

When to plant: Outdoors, in a seedframe, in late summer.
Temperature: 70° F.
Special treatment: Continual darkness, until the seeds sprout, produces the most successful germination.
Days to sprout: 10.
Light for seedlings: Half-day or more of sun.
Maturity time: Blooms in winter and spring of second year; some seedlings may bloom the first season if frost does not come until late in the fall.
Comments: *Viola* is easy to grow from seeds. For best performance, pick spent blooms before seeds form. In climates where hot summer weather tends to arrive early and suddenly upon the heels of spring, pansies and *Viola* will stay in bloom longer if they are afforded some shade and soil that is moist. Hybrid pansy and *Viola* seeds are listed by Burpee, Butcher, Park, Suttons, and Thompson & Morgan (*see also* Appendix).

WALLFLOWER: *See Cheiranthus*

5

Perennial Flowers to Grow from Seeds

There are sturdy perennial flowers—from the first dwarf *Iris* that flirts its velvety petals with the early spring snowfall to the greenish white Christmas rose (*Helleborus niger*) that really does bloom during warm spells in the middle of winter. While the ethereal beauty of flowers is not to be tallied up like a balance sheet, these hardy, herbaceous perennials do yield a big return—year after year—on a very small investment.

Herbaceous means that the stems are not persistently woody. Every spring they sprout from roots that live through winter in ground that is often cold, wet, and solidly frozen. The common peony, with its uncommonly beautiful flowers, is a perfect example. Dormant peony roots planted two inches deep in late summer send up bronzy green shoots—similar in appearance to asparagus spears—in the spring. These quickly expand into leathery, dark green foliage and sturdy stems topped by fragrant flowers. Thereafter, the foliage remains handsome until freezing in autumn; then they once more take to the underground until warm weather returns. But, like a truly great perennial, once planted in well-drained soil in a sunny place where it doesn't have to compete with greedy tree roots, the peony needs no particular attention—hardly ever needing to be dug and divided.

Apart from saving money, growing perennials from seeds makes sense because of the enormous variety not readily available in plant form. There are also numerous wild flowers that number among the perennials, some of which are on conservation lists and difficult to acquire legally unless you start them from seeds.

Individually potted seedlings of perennial flowers are more easily handled if they are grouped together in flats like this one which can be placed in a protected seedframe in the garden.

Although some perennials bloom during the first season, especially if they are given an early start indoors, most cannot be counted on for flowers until sometime during the second year. And a few, including the peony, may not bloom for five years.

The conditions under which perennial flower seeds sprout vary widely, depending on the individual plant. However, as a general rule, almost all of them, including the most difficult, are more likely to sprout readily in the spring than at any other time of the year.

ACHILLEA (yarrow; milfoil)

When to plant: Outdoors, in a protected place, in spring or summer; indoors, 8 to 10 weeks before planting-out time in the spring.
Temperature: 65°–75° F.
Special treatment: Seeds need light to sprout; sow on surface, and do not cover with planting medium.
Days to sprout: 10.
Light for seedlings: Half-day or more of sun, or start in a fluorescent-light garden (15–16 hours of light daily).
Maturity time: Yellow or rose and white flowers in the summer of the second season.
Comments: Easy to grow from seeds, *Achillea* tolerates drought. The flowers are prized for cutting, used fresh or dried in bouquets. The seeds are listed by Burpee, Butcher, Park, Suttons, and Thompson & Morgan (*see also* Appendix).

ACONITE: *See Aconitum*

ACONITUM (aconite; monkshood)

When to plant: Outdoors, in a seedframe, in late fall or early

winter; indoors, in early spring following refrigerator treat-
ment (see below).
Temperature: 60° F.
Special treatment: Seeds require freezing in order to sprout;
to start indoors, first place planting in freezer for 2 to 4
months.
Days to sprout: 20.
Light for seedlings: Half-day or more of sun, or start in a
fluorescent-light garden (15–16 hours of light daily).
Maturity time: First blue flowers bloom in autumn of the
second season.
Comments: A tall perennial, *Aconitum* is valued for fall bloom.
The seeds are listed by Butcher, Park, Suttons, and Thompson
& Morgan (*see also* Appendix).

ADAM'S NEEDLE: *See Yucca*

ADDER'S-TONGUE: *See Erythronium*

ADONIS (pheasant's-eye)

When to plant: Outdoors, in a seedframe, in late fall or early
spring; indoors, 6 to 8 weeks before planting-out time in the
spring.
Temperature: 60°–65° F.
Special treatment: None required.
Days to sprout: 10.
Light for seedlings: Half-day or more of sun, or start in a
fluorescent-light garden (15–16 hours of light daily).
Maturity time: Crimson, cup-shaped flowers in June of the
second season.
Comments: *Adonis* seeds are listed by Butcher, Park, Suttons,
and Thompson & Morgan (*see also* Appendix).

ALSTROEMERIA (lily-of-the-Incas; Peruvian lily)

When to plant: Outdoors, in a seedframe, in earliest spring.
Temperature: 60°–65° F.
Special treatment: None required.
Days to sprout: 15.
Light for seedlings: Half-day or more of sun.
Maturity time: Some bloom in the second season.
Comments: Where winter temperatures fall below 15° F., this perennial requires perfectly drained soil and a deep mulch. *Alstroemeria* is beautiful in the garden, but it is superb as a cut flower. The seeds are listed by Butcher, Park, Suttons, and Thompson & Morgan (*see also* Appendix).

ALYSSUM (madwort)

When to plant: Outdoors, in a seedframe, in earliest spring; indoors, 8 to 10 weeks before planting-out time in the spring.
Temperature: 65°–70° F.
Special treatment: None required.
Days to sprout: 10.
Light for seedlings: Half-day or more of sun, or start in a fluorescent-light garden (15–16 hours of light daily).
Maturity time: First blooms in the spring of the second season.
Comments: Seeds of perennial *Alyssum* are tiny, but easy to grow. They are listed by Burpee, Butcher, Park, Suttons, and Thompson & Morgan (*see also* Appendix).

ANAPHALIS (everlasting; life-everlasting)

When to plant: Outdoors, in a seedframe, in early spring or late fall.
Temperature: 60°–65° F.

Special treatment: None required.
Days to sprout: 10.
Light for seedlings: Half-day or more of sun.
Maturity time: First blooms often appear toward the end of the same season in which the seeds sprout.
Comments: This perennial is cultivated mostly for its flowers which easily dry for winter bouquets. The seeds are listed by Park and Thompson & Morgan (*see also* Appendix).

ANEMONE (windflower; lily-of-the-field)

When to plant: Outdoors, in a seedframe, in early spring or late fall.
Temperature: 60°–65° F.
Special treatment: None required.
Days to sprout: 15.
Light for seedlings: Half-day or more of sun.
Maturity time: First blooms in the second season.
Comments: Seeds of the spring-flowering *Anemone pulsatilla* are listed by Butcher, Park, Suttons, and Thompson & Morgan (*see also* Appendix); Park also lists *A. japonica*, one of the most beautiful of all hardy fall-flowering perennials.

ANTHEMIS (golden marguerite)

When to plant: Outdoors, in a seedframe, in spring or summer; indoors, 6 to 8 weeks before planting-out time in the spring.
Temperature: 65°–70° F.
Special treatment: None required.
Days to sprout: 8.
Light for seedlings: Half-day or more of sun, or start in a fluorescent-light garden (15–16 hours of light daily).
Maturity time: Yellow, daisylike flowers in the second season.
Comments: *Anthemis* is easy to grow; seeds are listed by

Burpee, Butcher, and Thompson & Morgan (*see also* Appendix).

AQUILEGIA (columbine)

When to plant: Outdoors, in a seedframe, in spring or summer; indoors, 6 to 8 weeks before planting-out time in the spring.
Temperature: 50°–60° F.
Special treatment: Seeds need light to sprout; sow on surface, and do not cover with planting medium.
Days to sprout: 15–30.
Light for seedlings: Half-day or more of sun, or start in a fluorescent-light garden (15–16 hours of light daily).
Maturity time: First blooms in the second season.
Comments: This perennial is easy to grow from seeds, and it is one of the most beautiful of all flowers for garden effect. Soil that is moist, well-drained, and rich in humus produces the best growth. The seeds are listed by Burpee, Butcher, Park, Suttons, and Thompson & Morgan (*see also* Appendix).

ARABIS (rock-cress)

When to plant: Outdoors, in a seedframe, in spring or summer; indoors, 6 to 8 weeks before planting-out time in the spring.
Temperature: 70° F.
Special treatment: Seeds need light to sprout; sow on surface, and do not cover with planting medium.
Days to sprout: 15.
Light for seedlings: Half-day or more of sun, or start in a fluorescent-light garden (15–16 hours of light daily).
Maturity time: First blooms in the second season.
Comments: Very easy to grow from seeds, *Arabis* blooms along with the spring bulbs and makes a wonderful edger,

carpeter, or cascading plant for the top of a wall. The seeds are listed by Burpee, Butcher, Park, Suttons, and Thompson & Morgan (*see also* Appendix).

ARENARIA (sandwort)

When to plant: Outdoors, in a seedframe, in early spring while the soil is cool.
Temperature: 60° F.
Special treatment: None required.
Days to sprout: 8.
Light for seedlings: Half-day or more of sun.
Maturity time: First blooms in the second season.
Comments: *Arenaria* is easy to grow from seeds; once established, it is tolerant of drought. Listed by Park and Suttons (*see also* Appendix).

ARMERIA (thrift; sea pink)

When to plant: Outdoors, in a seedframe, in spring or summer; indoors, 6 to 8 weeks before planting-out time in the spring.
Temperature: 70° F.
Special treatment: Germination is hastened by soaking the seeds in water at room temperature for 24 hours before sowing.
Days to sprout: 10
Light for seedlings: Half-day or more of sun, or start in a fluorescent-light garden (15–16 hours of light daily).
Maturity time: First blooms in the second season.
Comments: *Armeria* is easy to grow from seeds; once established, it is tolerant of drought. It is also excellent to grow in gardens near the seashore. The flowering season can be prolonged by clipping spent blooms before seeds form. *Armeria* is listed by Butcher, Park, Suttons, and Thompson & Morgan (*see also* Appendix).

ARNICA

When to plant: Outdoors, in a seedframe, in late fall or early spring.
Temperature: 60° F.
Special treatment: None required.
Days to sprout: 25.
Light for seedlings: Half-day or more of sun.
Maturity time: First blooms in the second season.
Comments: This drought-tolerant, low-growing daisy flower is easy to grow from seeds. It is listed by Park and Thompson & Morgan (*see also* Appendix).

ARTEMISIA (wormwood; old-woman; dusty miller)

When to plant: Outdoors, in a seedframe, in late fall or early spring.
Temperature: 65° F.
Special treatment: None required.
Days to sprout: 15.
Light for seedlings: Half-day or more of sun.
Maturity time: Interesting gray foliage plants by the end of the first season.
Comments: *Artemisia* seeds are listed by Park (*see also* Appendix).

ASPHODEL: *See Asphodelus*

ASPHODELUS (asphodel)

When to plant: Outdoors, in a seedframe, in spring or summer; indoors, 6 to 8 weeks before planting-out time in the spring.
Temperature: 70° F.
Special treatment: None required.

Days to sprout: 30.
Light for seedlings: Half-day or more of sun, or start in a fluorescent-light garden (15–16 hours of light daily).
Maturity time: First blooms in the second season.
Comments: This little-known perennial is easy to grow. Seeds are listed by Park (*see also* Appendix).

ASTER (Michaelmas daisy)

When to plant: Outdoors, in a seedframe, in early spring while the soil is cool.
Temperature: 60° F.
Special treatment: None required.
Days to sprout: 15.
Light for seedlings: Half-day or more of sun.
Maturity time: Some bloom in autumn of the first season.
Comments: *Asters* are available in various sizes, from 12 inches to 4 feet tall. They are among the best of all hardy perennials for flowers in late summer and fall, up to frost. The seeds are listed by Burpee, Butcher, Park, Suttons, and Thompson & Morgan (*see also* Appendix).

ASTILBE

When to plant: Outdoors, in a seedframe, in early spring while the soil is cool.
Temperature: 60° F.
Special treatment: None required.
Days to sprout: 25 or more.
Light for seedlings: Half-day of sun.
Maturity time: First blooms in the second season.
Comments: This hardy perennial needs well-drained, moist, humuslike soil in partial shade; it is an outstanding companion plant for the hardy lilies. *Astilbe* seeds are listed by Park and Thompson & Morgan (*see also* Appendix).

AUBRIETA

When to plant: Outdoors, in a seedframe, in spring or summer; indoors, 8 to 10 weeks before planting-out time in the spring.
Temperature: 65°–75° F.
Special treatment: None required.
Days to sprout: 20.
Light for seedlings: Half-day or more of sun, or start in a fluorescent-light garden (15–16 hours of light daily).
Maturity time: First blooms in the second season.
Comments: This hardy perennial is excellent in any sunny rock garden, or as an edger or bedder. The seeds are listed by Burpee, Butcher, Park, Suttons, and Thompson & Morgan (*see also* Appendix).

AVENS: *See Geum*

BABY'S-BREATH: *See Gypsophila*

BALLOONFLOWER: *See Platycodon*

BAPTISIA (false indigo)

When to plant: Outdoors, in a seedframe, in late fall or early spring.
Temperature: 60° F.
Special treatment: None required.
Days to sprout: 20.
Light for seedlings: Half-day or more of sun.
Maturity time: First blooms in the second or third season.
Comments: *Baptisia* is easy to grow; once established, this perennial is tolerant of drought. It grows about 3 feet tall and is valued for its blue flowers. The seeds are listed by Park and Thompson & Morgan (*see also* Appendix).

BEARDTONGUE: *See Penstemon*

BEE-BALM: *See Monarda*

BELAMCANDA (blackberry-lily)

When to plant: Outdoors, in a seedframe, in spring or summer; indoors, 8 to 12 weeks before planting-out time in the spring.
Temperature: 70° F.
Special treatment: None required.
Days to sprout: 15.
Light for seedlings: Half-day or more of sun, or start in a fluorescent-light garden (15–16 hours of light daily).
Maturity time: If seeds are started early indoors, as described above, *Belamcanda* blooms in August–September of the first season.
Comments: Blackberry-lily is one of the easiest and most rewarding of all hardy perennials to grow from seeds. In addition to fans of irislike foliage that offer interesting textural contrast to the garden, the late summer flowers are beautiful. Moreover, these flowers are followed by seedpods that open when ripe to reveal heads of blackberrylike seeds. These pods are excellent for use in dried arrangements. Seeds are listed by Park (*see also* Appendix).

BELLFLOWER: *See Campanula*

BERGENIA

When to plant: Outdoors, in a seedframe, in late fall or early spring.
Temperature: 60° F.
Special treatment: None required.
Days to sprout: 15.

Light for seedlings: Half-day of sun.

Maturity time: First blooms in the second season; attractive foliage in the first season.

Comments: This perennial has large evergreen leaves that are green spring–summer and crimson in cold weather. The flowers are rose pink. *Bergenia* needs soil that is moist, well-drained, and rich in humus; it is excellent to grow in partial shade near a stream or pond. The seeds are listed by Butcher, Park, and Thompson & Morgan (*see also* Appendix).

BETONY: *See Stachys*

BLACKBERRY-LILY: *See Belamcanda*

BLANKETFLOWER: *See Gaillardia*

BLAZING-STAR: *See Liatris*

BLEEDING-HEART: *See Dicentra*

BOUNCING BET: *See Saponaria*

BUTTON SNAKEROOT: *See Liatris*

CAMPANULA (bellflower)

When to plant: Outdoors, in a seedframe, in spring or summer; indoors, 8 to 12 weeks before planting-out time in the spring.

Temperature: 70° F.

Special treatment: None required.

Days to sprout: 20.

Light for seedlings: Half-day or more of sun, or start in a fluorescent-light garden (15–16 hours of light daily).

Maturity time: First blooms in the second season.

Comments: The perennial bellflowers are easy to grow from seeds, and they offer a variety of sizes (up to 5 feet tall) and habits (from trailing to upright). The flowers are usually white or in some shade of blue or amethyst violet; the bloom season may be any time from spring until late summer. The seeds are listed by Burpee, Butcher, Park, Suttons, and Thompson & Morgan (*see also* Appendix); Park and Thompson & Morgan offer the greatest variety.

CAMPION: *See Lychnis*

CAMPION: *See Silene*

CANDYTUFT: *See Iberis*

CARDINAL-FLOWER: *See Lobelia*

CARNATION: *See Dianthus*

CATANANCHE (cupid's-dart)

When to plant: Outdoors, in a seedframe, in early spring while the soil is cool.
Temperature: 50°–60° F.
Special treatment: None required.
Days to sprout: 20.
Light for seedlings: Half-day or more of sun.
Maturity time: May give some bloom in the first season.
Comments: Easy to grow from seeds, the violet blue flowers of this hardy perennial are excellent for garden effect as well as for cutting. The seeds are listed by Butcher, Park, Suttons, and Thompson & Morgan (*see also* Appendix).

CATCHFLY: *See Lychnis*

CATCHFLY: *See Silene*

CATMINT: *See Nepeta*

CATNIP: *See Nepeta*

CERASTIUM (snow-in-summer)

When to plant: Outdoors, in a seedframe, in early spring while the soil is cool.
Temperature: 65°–75° F.
Special treatment: None required.
Days to sprout: 15.
Light for seedlings: Half-day or more of sun.
Maturity time: Interesting silver foliage in the first season; white flowers beginning in the second season.
Comments: This hardy perennial is easy to grow from seeds and is excellent for use as an edger or in a rock garden. The seeds are listed by Burpee, Butcher, Park, Suttons, and Thompson & Morgan (*see also* Appendix).

CHINESE-LANTERN PLANT: *See Physalis*

CHRISTMAS-ROSE: *See Helleborus*

CHRYSANTHEMUM (pyrethrum; painted daisy; marguerite; Paris daisy; oxeye daisy; feverfew; dusty miller; Shasta daisy)

When to plant: Outdoors, in a seedframe, in spring or summer; indoors, 8 to 12 weeks before planting-out time in the spring.
Temperature: 60°–65° F.
Special treatment: None required.
Days to sprout: 10.
Light for seedlings: Half-day or more of sun, or start in a fluorescent-light garden (15–16 hours of light daily).

CHRYSANTHEMUM, (Ht. 5-6´)
AUTUMN GLORY (Full Sun)

Chrysanthemum *seedlings reach this size in about eight weeks, at which time they can be transplanted to the garden.*

Maturity time: If given an early start indoors, as described above, most species of *Chrysanthemum* will give some bloom toward the end of the first season.

Comments: Most species and varieties of *Chrysanthemum* are easy and rewarding to grow from seeds. In some catalogs you will find them listed by common names rather than according to the strict Latin designation of *Chrysanthemum*. The seeds are listed by Burpee, Butcher, Park, Sunnyslope Gardens, Suttons, and Thompson & Morgan (*see also* Appendix).

CINQUEFOIL: *See Potentilla*

CLEMATIS

When to plant: Outdoors, in a seedframe, in late fall or early spring.

Temperature: 60°–65° F.

Special treatment: None required.

Days to sprout: 50 or more.

Light for seedlings: Half-day or more of sun.

Maturity time: First blooms in the second season.

Comments: Not easy to grow from seeds, *Clematis* is a fascinating challenge. The seeds are listed by Butcher, Park, and Thompson & Morgan (*see also* Appendix); Thompson & Morgan offers the greatest variety.

COLUMBINE: *See Aquilegia*

CONEFLOWER: *See Rudbeckia*

CORAL-BELLS: *See Heuchera*

COREOPSIS (tickseed)

When to plant: Outdoors, in a seedframe, in spring or summer; indoors, 8 to 10 weeks before planting-out time in the spring.
Temperature: 65°–75° F.
Special treatment: Seeds need light to sprout; sow on surface, and do not cover with planting medium.
Days to sprout: 5.
Light for seedlings: Half-day or more of sun, or start in a fluorescent-light garden (15–16 hours of light daily).
Maturity time: First blooms in the second season; if seeds are started early indoors, as described above, some bloom may occur toward the end of the first summer.
Comments: Easy to grow from seeds, tickseed is an outstanding hardy perennial both for garden effect and for cut flowers. The seeds are listed by Burpee, Butcher, Park, Suttons, and Thompson & Morgan (*see also* Appendix).

CORONILLA (crown vetch)

When to plant: Outdoors in early spring while the soil is cool, preferably where they are to grow and mature.
Temperature: 60°–65° F.
Special treatment: None required.
Days to sprout: 10.
Light for seedlings: Half-day or more of sun.
Maturity time: First blooms in the second season.
Comments: Although an outstanding ground cover for roadsides and dry inclines, crown vetch is not recommended for small gardens. The seeds are listed by Burpee and Park (*see also* Appendix).

CORYDALIS

When to plant: Outdoors, in a seedframe, in late fall or early spring.

Temperature: 60°–65° F.
Special treatment: None required.
Days to sprout: 30.
Light for seedlings: Half-day of sun.
Maturity time: First blooms in second season.
Comments: This small-growing hardy perennial is excellent for a rock garden in partial shade, or for naturalizing with *Ferns, Trilliums, Hostas,* and bleeding-hearts in a nearly wild garden. The seeds are listed by Park and Thompson & Morgan (*see also* Appendix).

CRANESBILL: *See Geranium*

CROWN VETCH: *See Coronilla*

CUPID'S-DART: *See Catananche*

DAYLILY: *See Hemerocallis*

DELPHINIUM

When to plant: Outdoors, in a seedframe, in spring or summer; indoors, 8 to 10 weeks before planting-out time in the spring.
Temperature: 55°–65° F.
Special treatment: Generally speaking, none required, especially for the hybrids. Seeds of the species are more likely to germinate properly if the planting is first placed in a freezer for 2 to 4 months.
Days to sprout: 20.
Light for seedlings: Half-day or more of sun, or start in a fluorescent-light garden (15–16 hours of light daily).
Maturity time: If started early indoors, as described above, some of the hybrids will give some bloom toward the end of the first summer; all bloom in the second season.
Comments: For best results, *Delphiniums* need full sun and

rich soil that is moist and well-drained. Some may behave as annuals, others as biennials, and still others as hardy perennials. Many hybrid strains of *Delphinium* are listed in the catalogs of Burpee, Butcher, Park, Suttons, and Thompson & Morgan (*see also* Appendix).

DESERT-CANDLE: *See Eremurus*

DIANTHUS (pink; carnation; sweet William)

When to plant: Outdoors, in a seedframe, in spring or summer; indoors, 8 to 12 weeks before planting-out time in the spring.
Temperature: 65°–75° F.
Special treatment: None required.
Days to sprout: 5.
Light for seedlings: Half-day or more of sun, or start in a fluorescent-light garden (15–16 hours of light daily).
Maturity time: With the exception of the sweet William types (which usually do not bloom until the second season), most *Dianthus* will give some flowers toward midsummer of the first season, provided they are given an early start indoors, as described above.
Comments: *Dianthus*, like the genus *Chrysanthemum*, offers enormous variety for a flower garden that receives a half-day or more of sun. All of them are also excellent for cultivation in container gardens outdoors. The carnations may also be potted and kept during the winter for flowers in a greenhouse that is sunny, airy, and moist. The seeds are listed by Burpee, Butcher, Park, Suttons, and Thompson & Morgan (*see also* Appendix).

DICENTRA (Dutchman's breeches; bleeding-heart)

When to plant: Outdoors, in a seedframe, in late fall or early

Picotee Dianthus *like this one are easy to grow from seeds; if started early indoors, some blooms may appear toward the end of the first summer.*

winter; indoors, in early spring following refrigerator treatment (see below).

Temperature: 60° F.

Special treatment: Seeds require freezing in order to sprout; to start indoors, first place planting in freezer for 2 to 4 months.

Days to sprout: 50 or more.

Light for seedlings: Half-day of sun, or start in a fluorescent-light garden (15–16 hours of light daily).

Maturity time: First blooms in the second season.

Comments: Not easy to grow from seeds, *Dicentra* is a fascinating challenge. The seeds are listed by Butcher and Park (*see also* Appendix).

DICTAMNUS (gasplant; fraxinella)

When to plant: Outdoors, in a seedframe, in late fall or early winter; indoors, in early spring following refrigerator treatment (see below).

Temperature: 60° F.

Special treatment: Seeds require freezing in order to sprout; to start indoors, first place planting in freezer for 2 to 4 months.

Days to sprout: 50 or more.

Light for seedlings: Half-day of sun or more, or start in a fluorescent-light garden (15–16 hours of light daily).

Maturity time: First blooms in the third season.

Comments: Although *Dictamnus* is not easy to grow from seeds, once established, it is a tough, long-lived perennial. The seeds are listed by Butcher, Park, and Thompson & Morgan (*see also* Appendix).

DODECATHEON (shooting-star)

When to plant: Outdoors, in a seedframe, in late fall or early

winter; indoors, in early spring following refrigerator treatment (see below).

Temperature: 60° F.

Special treatment: Seeds require freezing in order to sprout; to start indoors, first place planting in freezer for 2 to 4 months.

Days to sprout: 30 or more.

Light for seedlings: Half-day of sun, or start in a fluorescent-light garden (15–16 hours of light daily).

Maturity time: First blooms in second season.

Comments: Not easy to grow from seeds, *Dodecatheon* is beautiful enough to make the challenge worthwhile. This hardy perennial is ideal for planting in a rock or wild garden in soil that is rich in humus, moist, and well-drained. Shade from hottest sun is also needed. *Dodecatheon* seeds are listed by Park and Thompson & Morgan (*see also* Appendix).

DOGTOOTH VIOLET: *See Erythronium*

DORONICUM (leopardsbane)

When to plant: Outdoors, in a seedframe, in spring or summer; indoors, 8 to 10 weeks before planting-out time in the spring.

Temperature: 65°–75° F.

Special treatment: None required.

Days to sprout: 15.

Light for seedlings: Half-day or more of sun, or start in a fluorescent-light garden (15–16 hours of light daily).

Maturity time: First blooms in the spring of the second season.

Comments: Easy to grow from seeds, *Doronicum* is an ideal companion for spring-flowering bulbs and is excellent for cutting. The seeds are listed by Butcher, Park, and Thompson & Morgan (*see also* Appendix).

DRYAS (mountain avens)

When to plant: Outdoors, in a seedframe, in late fall or early winter; indoors, in early spring following refrigerator treatment (see below).
Temperature: 60° F.
Special treatment: Seeds require freezing in order to sprout; to start indoors, first place planting in freezer for 2 to 4 months.
Days to sprout: 50 or more.
Light for seedlings: Half-day or more of sun, or start in a fluorescent-light garden (15–16 hours of light daily).
Maturity time: First blooms in second season.
Comments: *Dryas* is an ideal hardy perennial for planting in a sunny rock garden. The seeds are listed by Park and Thompson & Morgan (*see also* Appendix).

DUSTY MILLER: *See Artemisia*

DUSTY MILLER: *See Chrysanthemum*

DUTCHMAN'S BREECHES: *See Dicentra*

ECHINACEA (purple coneflower)

When to plant: Outdoors, in a seedframe, in spring or summer; indoors, 8 to 10 weeks before planting-out time in the spring.
Temperature: 70° F.
Special treatment: None required.
Days to sprout: 10 to 25.
Light for seedlings: Half-day or more of sun, or start in a fluorescent-light garden (15–16 hours of light daily).
Maturity time: First blooms in the second season.
Comments: This hardy perennial grows to 3 feet tall. In late summer it sends up many long-stemmed, reddish purple

daisy flowers with a prominent cone in the center of each. *Echinacea* is excellent for garden effect and cutting. The seeds are listed by Butcher, Park, and Thompson & Morgan (*see also* Appendix).

ECHINOPS (globe-thistle)

When to plant: Outdoors, in a seedframe, in spring or summer; indoors, 8 to 10 weeks before planting-out time in the spring.
Temperature: 65°–75° F.
Special treatment: None required.
Days to sprout: 15.
Light for seedlings: Half-day or more of sun, or start in a fluorescent-light garden (15–16 hours of light daily).
Maturity time: First blooms in the second season.
Comments: Easy to grow from seeds, this hardy perennial sends up silvery blue, globe-shaped flower heads from mid-summer to late summer. It is outstanding and belongs in almost every sunny, perennial flower border. The seeds are listed by Butcher, Park, Suttons, and Thompson & Morgan (*see also* Appendix).

EDELWEISS: *See Leontopodium*

EREMURUS (foxtail lily; desert-candle)

When to plant: Outdoors, in a seedframe, in late fall or early winter; indoors, in early spring following refrigerator treatment (see below).
Temperature: 60° F.
Special treatment: Seeds require freezing in order to sprout; to start indoors, first place planting in freezer for 2 to 4 months.
Days to sprout: 30.

Light for seedlings: Half-day or more of sun, or start in a fluorescent-light garden (15–16 hours of light daily).
Maturity time: First blooms in the second season.
Comments: In rich, moist, well-drained soil, *Eremurus* is potentially one of the most beautiful of all hardy perennials. The seeds are listed by Butcher, Park, and Thompson & Morgan (*see also* Appendix).

ERIGERON (fleabane)

When to plant: Outdoors, in a seedframe, in late fall or early spring.
Temperature: 60° F.
Special treatment: None required.
Days to sprout: 15.
Light for seedlings: Half-day or more of sun.
Maturity time: First blooms in the second season; occasionally toward the end of the first.
Comments: Easy to grow from seeds, this hardy perennial is prized for its compact habit and its orange, pink, blue, or white daisy flowers. It is good for garden effect or cutting. The seeds are listed by Butcher, Park, Suttons, and Thompson & Morgan (*see also* Appendix).

ERINUS

When to plant: Outdoors, in a seedframe, in spring or summer; indoors, 8 to 10 weeks before planting-out time in the spring.
Temperature: 70° F.
Special treatment: None required.
Days to sprout: 25.
Light for seedlings: Half-day or more of sun, or start in a fluorescent-light garden (15–16 hours of light daily).
Maturity time: First blooms in the second season.
Comments: This little-known perennial is easy to grow from

seeds. *Erinus* stands about 6 inches tall and produces pink, red, or purple flowers. The seeds are listed by Butcher, Park, and Thompson & Morgan (*see also* Appendix).

ERYNGIUM (eryngo; sea-holly)

When to plant: Outdoors, in a seedframe, in late fall or early winter; indoors, in early spring following refrigerator treatment (see below).
Temperature: 60° F.
Special treatment: Seeds require freezing in order to sprout; to start indoors, first place planting in freezer for 2 to 4 months.
Days to sprout: 50 or more.
Light for seedlings: Half-day or more of sun, or start in a fluorescent-light garden (15–16 hours of light daily).
Maturity time: First blooms in the second season.
Comments: This hardy perennial tolerates drought and also is choice for gardens near the seashore. The seeds are listed by Butcher, Park, Suttons, and Thompson & Morgan (*see also* Appendix).

ERYNGO: *See Eryngium*

ERYTHRONIUM (adder's-tongue; dogtooth violet; trout-lily; fawn-lily)

When to plant: Outdoors, in a seedframe, in late fall or early winter; indoors, in early spring following refrigerator treatment (see below).
Temperature: 60° F.
Special treatment: Seeds require freezing in order to sprout; to start indoors, first place planting in freezer for 2 to 4 months.
Days to sprout: 50 or more.

Light for seedlings: Half-day of sun, or start in a fluorescent-light garden (15–16 hours of light daily).
Maturity time: First blooms in the third season.
Comments: This North American native wild flower is a choice plant for growing in a partly shaded site where the soil is rich in humus, moist, and well-drained. It is a good companion for hardy ferns, *Trilliums*, *Hosta*, *Astilbe*, and native lilies. The seeds are listed by Park and Thompson & Morgan (*see also* Appendix).

EUPHORBIA

When to plant: Outdoors, in a seedframe, in early spring while the soil is cool.
Temperature: 60°–65° F.
Special treatment: None required.
Days to sprout: 15.
Light for seedlings: Half-day or more of sun.
Maturity time: Interesting foliage plants the first year; flowers beginning in the second season.
Comments: The best-known members of the genus *Euphorbia*—Christmas poinsettia and crown-of-thorns—bear no obvious resemblance to the hardy perennial species. Once established, hardy perennial *Euphorbia* plants are tolerant of drought. Seeds are listed by Butcher, Park, and Thompson & Morgan (*see also* Appendix).

EVERLASTING: *See Anaphalis*

FALSE INDIGO: *See Baptisia*

FAWN-LILY *See Erythronium*

FEVERFEW: *See Chrysanthemum*

FLAG: *See Iris*

FLAX: *See Linum*

FLEABANE: *See Erigeron*

FLEUR-DE-LIS: *See Iris*

FOXTAIL LILY: *See Eremurus*

FRAXINELLA: *See Dictamnus*

FUNKIA: *See Hosta*

GAILLARDIA (blanketflower)

When to plant: Outdoors, in a seedframe, in spring or summer; indoors, 8 to 10 weeks before planting-out time in the spring.
Temperature: 65°–75° F.
Special treatment: None required.
Days to sprout: 20.
Light for seedlings: Half-day or more of sun, or start in a fluorescent-light garden (15–16 hours of light daily).
Maturity time: If started early indoors, as described above, some bloom toward the end of the first season.
Comments: *Gaillardia* is an extremely easy perennial to grow from seeds; once established, it is tolerant of drought. Blanketflowers are excellent for garden effect and cutting. The seeds are listed by Burpee, Butcher, Park, Suttons, and Thompson & Morgan (*see also* Appendix). Note: Some *Gaillardia* are annuals, others are hardy perennials, as indicated by catalog descriptions.

GASPLANT: *See Dictamnus*

GAYFEATHER: *See Liatris*

GENTIAN: *See Gentiana*

GENTIANA (gentian)

When to plant: Outdoors, in a seedframe, in late fall or early winter; indoors, in early spring following refrigerator treatment (see below).
Temperature: 60° F.
Special treatment: Seeds require freezing in order to sprout; to start indoors, first place planting in freezer for 2 to 4 months.
Days to sprout: 30.
Light for seedlings: Half-day or more of sun, or start in a fluorescent-light garden (15–16 hours of light daily).
Maturity time: First blooms in the second season.
Comments: This hardy perennial—treasured for its blue, blue flowers—is not easy to grow from seeds, but offers a worthwhile challenge. It is listed by Butcher, Park, Suttons, and Thompson & Morgan (*see also* Appendix).

GERANIUM (cranesbill)

When to plant: Outdoors, in a seedframe, in late fall or early spring.
Temperature: 60° F.
Special treatment: None required.
Days to sprout: 20 or more.
Light for seedlings: Half-day or more of sun.
Maturity time: First blooms in the second season.
Comments: The true *Geranium* is a hardy perennial; the geranium of florists is a member of a related genus known as *Pelargonium* and is a tender perennial (treated in this book in

Chapter 3 as an annual). The true *Geranium* is excellent as a border or for planting in a nearly wild garden in sun or partial shade. The seeds are listed by Park and Thompson & Morgan (*see also* Appendix).

GERMANDER: *See Teucrium*

GEUM (avens)

When to plant: Outdoors, in a seedframe, in spring or summer; indoors, 8 to 10 weeks before planting-out time in the spring.
Temperature: 70°–75° F.
Special treatment: None required.
Days to sprout: 25.
Light for seedlings: Half-day or more of sun, or start in a fluorescent-light garden (15–16 hours of light daily).
Maturity time: If started early indoors, as described above, may give some bloom toward the end of the first season.
Comments: This hardy perennial is easy to grow from seeds. It is excellent for garden effect and cutting. If old flowers are picked before seeds form, the flowering season will extend over a long period. *Geum* seeds are listed by Burpee, Butcher, Park, Suttons, and Thompson & Morgan (*see also* Appendix).

GLOBEFLOWER: *See Trollius*

GLOBE-THISTLE: *See Echinops*

GOLDEN MARGUERITE: *See Anthemis*

GREEK VALERIAN: *See Polemonium*

GYPSOPHILA (baby's-breath)

When to plant: Outdoors, in a seedframe, in spring or summer; indoors, 8 to 10 weeks before planting-out time in the spring.
Temperature: 70°–75° F.
Special treatment: None required.
Days to sprout: 10.
Light for seedlings: Half-day or more of sun, or start in a fluorescent-light garden (15–16 hours of light daily).
Maturity time: First blooms in the second season; if started early indoors, as described above, may bloom toward the end of the first season.
Comments: An easy hardy perennial to grow from seeds for garden effect and for cutting, baby's breath can be used, fresh or dried, in bouquets. The seeds are listed by Burpee, Butcher, Park, Suttons, and Thompson & Morgan (*see also* Appendix).

HELENIUM (sneezeweed)

When to plant: Outdoors, in a seedframe, in spring or summer; indoors, 8 to 10 weeks before planting-out time in the spring.
Temperature: 70° F.
Special treatment: None required.
Days to sprout: 5.
Light for seedlings: Half-day or more of sun, or start in a fluorescent-light garden (15–16 hours of light daily).
Maturity time: First blooms in the second season.
Comments: This hardy perennial is easy to grow from seeds. It grows to 4 feet tall, forming bushes that are literally covered with daisy flowers of many colors in late summer. The seeds are listed by Butcher, Park, and Thompson & Morgan (*see also* Appendix).

HELIOPSIS (oxeye)

When to plant: Outdoors, in a seedframe, in spring or summer; indoors, 8 to 10 weeks before planting-out time in the spring.
Temperature: 70° F.
Special treatment: None required.
Days to sprout: 10.
Light for seedlings: Half-day or more of sun, or start in a fluorescent-light garden (15–16 hours of light daily).
Maturity time: First blooms in the second season.
Comments: This hardy perennial is easy to grow from seeds and is excellent for planting toward the back of a border. The seeds are listed by Butcher, Park, and Thompson & Morgan (*see also* Appendix).

HELLEBORUS (Christmas-rose; Lenten-rose)

When to plant: Outdoors, in a seedframe, in late fall or early winter; indoors, in early spring following refrigerator treatment (see below).
Temperature: 60° F.
Special treatment: Seeds require freezing in order to sprout; to start indoors, first place planting in freezer for 2 to 4 months.
Days to sprout: 30 or more.
Light for seedlings: Half-day of sun, or start in a fluorescent-light garden (15–16 hours of light daily).
Maturity time: First blooms in the third season.
Comments: These are the only commonly cultivated, hardy perennials that bloom outdoors during periods of unseasonably warm weather in winter and early spring. *Helleborus* needs moist soil that is rich in humus and well-drained; it should be planted in a site that is partly shaded in summer. The seeds are listed by Butcher, Park, Suttons, and Thompson & Morgan (*see also* Appendix).

HEMEROCALLIS (daylily)

When to plant: Outdoors, in a seedframe, in late fall or early spring.
Temperature: 60° F.
Special treatment: None required.
Days to sprout: 15.
Light for seedlings: Half-day or more of sun.
Maturity time: First blooms may appear in the second season, but more likely in the third.
Comments: Easy to grow from seeds, daylily is one of the best of all hardy perennials. The seeds are listed by Burpee, Park, and Thompson & Morgan (*see also* Appendix).

HERNIARIA (herniary)

When to plant: Outdoors, in a seedframe, in early spring while the soil is cool.
Temperature: 60°–65° F.
Special treatment: None required.
Days to sprout: 10.
Light for seedlings: Half-day or more of sun.
Maturity time: First blooms in the second season.
Comments: This little-known carpeter or trailing plant is easy to grow and deserves to be more widely cultivated. The seeds are listed by Park and Thompson & Morgan (*see also* Appendix).

HERNIARY: *See Herniaria*

HEUCHERA (coral-bells)

When to plant: Outdoors, in a seedframe, in late fall or early spring.

Temperature: 65° F.
Special treatment: Seeds need light to sprout; sow on surface, and do not cover with planting medium.
Days to sprout: 10.
Light for seedlings: Up to a half-day of sun.
Maturity time: First blooms in the second season.
Comments: A favorite among all perennials, coral-bells are ideal as an edger or in clumps among other flowers. The seeds are listed by Burpee, Butcher, Park, Suttons, and Thompson & Morgan (*see also* Appendix).

HIBISCUS (rosemallow)

When to plant: Outdoors, in a seedframe, in spring or summer; indoors, 8 to 10 weeks before planting-out time in the spring.
Temperature: 70° F.
Special treatment: None required.
Days to sprout: 15 or more.
Light for seedlings: Half-day or more of sun, or start in a fluorescent-light garden (15–16 hours of light daily).
Maturity time: If started early indoors, as described above, first blooms toward the end of the first season.
Comments: Easy to grow from seeds, *Hibiscus* has spectacular flowers on plants that grow to 4 feet tall. The seeds are listed by Park (*see also* Appendix).

HOSTA (plantain-lily; funkia)

When to plant: Outdoors, in a seedframe, in spring or summer; indoors, 8 to 10 weeks before planting-out time in the spring.
Temperature: 70° F.
Special treatment: None required.
Days to sprout: 15 or more.

Light for seedlings: Half-day of sun, or start in a fluorescent-light garden (15–16 hours of light daily).

Maturity time: Interesting foliage plants in the first season, flowers the second.

Comments: One of the best of all hardy perennials for growing in partly shaded areas, *Hosta* needs rich, moist, well-drained soil. The seeds are listed by Park and Thompson & Morgan (*see also* Appendix).

HOUSELEEK: *See Sempervivum*

HYPERICUM (St.-John's-wort)

When to plant: Outdoors, in a seedframe, in spring or summer; indoors, 8 to 10 weeks before planting-out time in the spring.

Temperature: 70° F.

Special treatment: None required.

Days to sprout: 20.

Light for seedlings: Half-day or more of sun, or start in a fluorescent-light garden (15–16 hours of light daily).

Maturity time: First blooms in the second season.

Comments: *Hypericum* is one of the best of the hardy perennials. Shrublike bushes, to 3 feet tall, produce a long season of bloom; the flowers are clear bright yellow. *Hypericum* seeds are listed by Park (*see also* Appendix).

IBERIS (candytuft)

When to plant: Outdoors, in a seedframe, in spring or summer; indoors, 8 to 10 weeks before planting-out time in the spring.

Temperature: 70° F.

Special treatment: None required.

Days to sprout: 20.

Light for seedlings: Half-day or more of sun, or start in a fluorescent-light garden (15–16 hours of light daily).

Maturity time: First blooms in the second season.

Comments: Perennial *Iberis sempervirens*, or evergreen candytuft, makes an ideal edger which can also be clipped to give the effect of a very low hedge. The seeds are listed by Burpee, Butcher, Park, Suttons, and Thompson & Morgan (*see also* Appendix).

IRIS (flag; fleur-de-lis)

When to plant: Outdoors, in a seedframe, in late fall or early winter; indoors, in early spring following refrigerator treatment (see below).

Temperature: 60° F.

Special treatment: Seeds require freezing in order to sprout; to start indoors, first place planting in freezer for 2 to 4 months.

Days to sprout: 20 or more.

Light for seedlings: Half-day or more of sun, or start in a fluorescent-light garden (15–16 hours of light daily).

Maturity time: First blooms occasionally in second season, usually in the third.

Comments: Besides the more common German (or bearded) iris, it is also possible to grow the choice Japanese, *spuria*, and Siberian types from seeds. Various kinds of *Iris* seeds are listed by Butcher, Park, and Thompson & Morgan (*see also* Appendix).

IRISH MOSS: *See Sagina*

JACOB'S-LADDER: *See Polemonium*

JASIONE

When to plant: Outdoors, in a seedframe, in spring or sum-

mer; indoors, 8 to 10 weeks before planting-out time in the spring.
Temperature: 70° F.
Special treatment: None required.
Days to sprout: 25.
Light for seedlings: Half-day or more of sun, or start in a fluorescent-light garden (15–16 hours of light daily).
Maturity time: First blooms in the second season.
Comments: This little-known perennial grows 6 to 12 inches tall and gives a long season of bloom. The flowers are tiny, sky blue bells. *Jasione* seeds are listed by Park and Thompson & Morgan (*see also* Appendix).

KNIPHOFIA (torch-lily; red hot poker; tritoma)

When to plant: Outdoors, in a seedframe, in spring or summer; indoors, 8 to 10 weeks before planting-out time in the spring.
Temperature: 65°–75° F.
Special treatment: None required.
Days to sprout: 20.
Light for seedlings: Half-day or more of sun, or start in a fluorescent-light garden (15–16 hours of light daily).
Maturity time: First blooms in the second season.
Comments: Easy to grow from seeds, *Kniphofia* (which is often listed in catalogs as tritoma) produces a great effect in the garden and is also outstanding as a cut flower. The seeds are listed by Burpee, Butcher, Park, Suttons, and Thompson & Morgan (*see also* Appendix).

LAMB'S-EARS: *See Stachys*

LAVANDULA (lavender)

When to plant: Outdoors, in a seedframe, in spring or sum-

Kniphofia *is an easy perennial to grow from seeds. It produces flowers in the summer of the second season and is excellent for cutting.* Courtesy George W. Park Seed Co., Inc.

mer; indoors, 8 to 10 weeks before planting-out time in the spring.

Temperature: 65°–75° F.

Special treatment: None required.

Days to sprout: 15.

Light for seedlings: Half-day or more of sun, or start in a fluorescent-light garden (15–16 hours of light daily).

Maturity time: First blooms in the second season.

Comments: This hardy perennial herb belongs in almost any flower border or garden that is sunny. Provide moist, well-drained soil. The seeds are listed by Burpee, Butcher, Park, Suttons, and Thompson & Morgan (*see also* Appendix).

LAVENDER: *See Lavandula*

LAVENDER-COTTON: *See Santolina*

LENTEN-ROSE: *See Helleborus*

LEONTOPODIUM (edelweiss)

When to plant: Outdoors, in a seedframe, in late fall or early spring.

Temperature: 60° F.

Special treatment: Seeds need light to sprout; sow on surface, and do not cover with planting medium.

Days to sprout: 10.

Light for seedlings: Half-day or more of sun.

Maturity time: First blooms in the second season.

Comments: This hardy perennial is excellent for a sunny rock garden; once established, it tolerates drought. The seeds are listed by Butcher, Park, and Thompson & Morgan (*see also* Appendix).

LEOPARDSBANE: *See Doronicum*

LEWISIA

When to plant: Outdoors, in a seedframe, in late fall or early winter; indoors, in early spring following refrigerator treatment (see below).
Temperature: 60° F.
Special treatment: Seeds require freezing in order to sprout; to start indoors, first place planting in freezer for 2 to 4 months.
Days to sprout: 30 or more.
Light for seedlings: Half-day or more of sun, or start in a fluorescent-light garden (15–16 hours of light daily).
Maturity time: First blooms in the third season.
Comments: Once established, this hardy perennial tolerates drought. The seeds are listed by Park (*see also* Appendix).

LIATRIS (blazing-star; button snakeroot; gayfeather)

When to plant: Outdoors, in a seedframe, in late fall or early spring.
Temperature: 65°–70° F.
Special treatment: None required.
Days to sprout: 20.
Light for seedlings: Half-day or more of sun.
Maturity time: First blooms in the second season.
Comments: Easy to grow from seeds, this North American native is great for garden effect and for cutting. Once established, *Liatris* is very tolerant of drought. The seeds are listed by Butcher, Park, Suttons, and Thompson & Morgan (*see also* Appendix).

LIFE-EVERLASTING: *See Anaphalis*

LILY-OF-THE-FIELD: *See Anemone*

LILY-OF-THE-INCAS: *See Alstroemeria*

LILY-TURF: *See Liriope*

LINUM (flax)

When to plant: Outdoors, in a seedframe, in spring or summer; indoors, 8 to 10 weeks before planting-out time in the spring.
Temperature: 70° F.
Special treatment: None required.
Days to sprout: 25.
Light for seedlings: Half-day or more of sun, or start in a fluorescent-light garden (15–16 hours of light daily).
Maturity time: First blooms in the second season.
Comments: Perennial flax is a choice flowering plant for a border or rock garden. Seeds are listed by Burpee, Butcher, Park, Suttons, and Thompson & Morgan (*see also* Appendix).

LIRIOPE (lily-turf)

When to plant: Outdoors, in a seedframe, in late fall or early spring.
Temperature: 65°–70° F.
Special treatment: Soak the seeds in water at room temperature for 24 hours before planting.
Days to sprout: 30.
Light for seedlings: Up to a half-day of sun.
Maturity time: First blooms in the second or third season.
Comments: Lily-turf is a superb ground cover in mostly shaded areas; it needs rich, moist, well-drained soil. The seeds are listed by Park (*see also* Appendix).

LIVE FOREVER: *See Sedum*

LIVE FOREVER: *See Sempervivum*

LOBELIA (cardinal-flower)

When to plant: Outdoors, in a seedframe, in late fall or early winter; indoors, in early spring following refrigerator treatment (see below).
Temperature: 60° F.
Special treatment: Seeds require freezing in order to sprout; to start indoors, first place planting in freezer for 2 to 4 months. *Lobelia* seeds also need light to sprout; sow on surface, and do not cover with planting medium.
Days to sprout: 20.
Light for seedlings: Half-day or more of sun, or start in a fluorescent-light garden (15–16 hours of light daily).
Maturity time: First blooms in the second season.
Comments: *Lobelia*, a hardy perennial, needs rich, moist, well-drained soil; it generally does best in partial shade. Some have vivid red flowers, others are blue. The seeds are listed by Butcher, Park, and Thompson & Morgan (*see also* Appendix).

LOOSESTRIFE: *See Lythrum*

LUPINE: *See Lupinus*

LUPINUS (lupine)

When to plant: Outdoors, where they are to grow and bloom, in early spring.
Temperature: 70° F.
Special treatment: Soak the seeds in water at room temperature for 24 hours before planting.
Days to sprout: 20.
Light for seedlings: Half-day or more of sun.
Maturity time: First blooms in the second season.
Comments: Lupines are easy to grow from seeds, but they cannot withstand transplanting. Provide moist, well-drained

soil and a sunny site. The seeds are listed by Burpee, Butcher, Park, Suttons, and Thompson & Morgan (*see also* Appendix).

LYCHNIS (campion; catchfly)

When to plant: Outdoors, in a seedframe, in spring or summer; indoors, 8 to 10 weeks before planting-out time in the spring.
Temperature: 70° F.
Special treatment: Seeds need light to sprout; sow on surface, and do not cover with planting medium.
Days to sprout: 25.
Light for seedlings: Half-day or more of sun, or start in a fluorescent-light garden (15–16 hours of light daily).
Maturity time: First blooms in the second season.
Comments: *Lychnis* is a hardy perennial that is easy to grow from seeds; once established, it is tolerant of drought. It is excellent for garden effect as well as for cutting. The seeds are listed by Burpee, Butcher, Park, Suttons, and Thompson & Morgan (*see also* Appendix).

LYTHRUM (loosestrife)

When to plant: Outdoors, in a seedframe, in late fall or early spring.
Temperature: 65° F.
Special treatment: None required.
Days to sprout: 15.
Light for seedlings: Half-day or more of sun.
Maturity time: First blooms in the second season.
Comments: This hardy perennial is easy to grow from seeds; it needs a sunny site that is rich in humus, moist, and well-drained. The seeds are listed by Butcher, Park, and Thompson & Morgan (*see also* Appendix).

MADWORT: *See Alyssum*

MARGUERITE: *See Chrysanthemum*

MEADOW-RUE: *See Thalictrum*

MECONOPSIS (Welsh-poppy)

When to plant: Outdoors, in a seedframe, in late fall or early spring.
Temperature: 65° F.
Special treatment: None required.
Days to sprout: 20.
Light for seedlings: Half-day or more of sun.
Maturity time: First blooms in the second or third season.
Comments: Although it is not the easiest of perennials to grow, *Meconopsis* is one of the most beautiful for garden effect. It needs soil that is rich in humus, moist, and well-drained. The seeds are listed by Park, Suttons, and Thompson & Morgan (*see also* Appendix).

MICHAELMAS DAISY: *See Aster*

MILFOIL: *See Achillea*

MONARDA (bee-balm; Oswego tea)

When to plant: Outdoors, in a seedframe, in spring or summer; indoors, 8 to 10 weeks before planting-out time in the spring.
Temperature: 70° F.
Special treatment: None required.
Days to sprout: 15.
Light for seedlings: Half-day or more of sun.

Meconopsis is not easy to grow from seeds, but many gardeners accept the challenge because of its extraordinarily beautiful, blue poppy flowers. Courtesy the George W. Park Seed Co., Inc.

Maturity time: First blooms in the second season.
Comments: Easy to grow from seeds, *Monarda* makes a valuable plant for almost any flower garden. The leaves have a pleasing mint scent. The seeds are listed by Butcher, Park, and Thompson & Morgan (*see also* Appendix).

MONKSHOOD: *See Aconitum*

MOUNTAIN AVENS: *See Dryas*

NEPETA (catmint; catnip)

When to plant: Outdoors, in a seedframe, in spring or summer; indoors, 8 to 10 weeks before planting-out time in the spring.
Temperature: 70° F.
Special treatment: None required.
Days to sprout: 10.
Light for seedlings: Half-day or more of sun, or start in a fluorescent-light garden (15–16 hours of light daily).
Maturity time: First blooms in the second season.
Comments: Extremely easy to grow, *Nepeta* is ideal for edging or to plant where the trailing stems can clamber over a rock wall. The seeds are listed by Butcher, Park, Suttons, and Thompson & Morgan (*see also* Appendix).

OBEDIENCE PLANT: *See Physostegia*

OLD-WOMAN: *See Artemisia*

ORIENTAL POPPY: *See Papaver*

OSWEGO TEA: *See Monarda*

OXEYE: *See Heliopsis*

OXEYE DAISY: *See Chrysanthemum*

PAEONIA (peony)

When to plant: Outdoors, in a seedframe, in late fall or early winter; indoors, in early spring following refrigerator treatment (see below).
Temperature: 60° F.
Special treatment: Seeds require freezing in order to sprout; to start indoors, first place planting in freezer for 2 to 4 months.
Days to sprout: 50 or more.
Light for seedlings: Half-day or more of sun, or start in a fluorescent-light garden (15–16 hours of light daily).
Maturity time: First blooms in 4 to 6 years.
Comments: Growing peonies from seeds requires more patience than anything else; however, after waiting for several years, the opening of the first flowers is cause for celebration. Peony seeds are listed by Park and Thompson & Morgan (*see also* Appendix).

PAINTED DAISY: *See Chrysanthemum*

PAPAVER (Oriental poppy)

When to plant: Outdoors, in a seedframe, in spring or summer.
Temperature: 70° F.
Special treatment: Continual darkness, until the seeds sprout, produces the most successful germination.
Days to sprout: 14.
Light for seedlings: Half-day or more of sun.
Maturity time: First blooms in the second or third season.
Comments: Oriental poppy seedlings are difficult to transplant; for best results, let them grow and bloom in the seed-

frame, then transplant to permanent position while the plants are dormant in late summer. The seeds are listed by Butcher, Park, Suttons, and Thompson & Morgan (*see also* Appendix).

PARIS DAISY: *See Chrysanthemum*

PEARLWORT: *See Sagina*

PENSTEMON (beardtongue)

When to plant: Outdoors, in a seedframe, in late fall or early spring.
Temperature: 65°–70° F.
Special treatment: None required.
Days to sprout: 10.
Light for seedlings: Half-day or more of sun.
Maturity time: First blooms in the second season.
Comments: Once established, *Penstemon* is tolerant of drought. This North American native is available in an infinite variety, both in natural species and hybrids. For best results in the garden, *Penstemon* needs moist, well-drained soil and full sun. The seeds are listed by Park, Suttons, and Thompson & Morgan (*see also* Appendix).

PEONY: *See Paeonia*

PERUVIAN LILY: *See Alstroemeria*

PHEASANT'S-EYE: *See Adonis*

PHLOX

When to plant: Outdoors, in a seedframe, in late fall or early

winter; indoors, in early spring following refrigerator treatment (see below).

Temperature: 65°–70° F.

Special treatment: Seeds require freezing in order to sprout; to start indoors, first place planting in freezer for 2 to 4 months.

Days to sprout: 25 or more.

Light for seedlings: Half-day or more of sun, or start in a fluorescent-light garden (15–16 hours of light daily).

Maturity time: First blooms in the second season.

Comments: *Phlox* is not the easiest of perennials to grow from seeds, but the results can be very interesting. Once the seedlings have bloomed, pull out and discard any with flowers that have muddy or unattractive colors. The seeds are listed by Butcher, Park, and Thompson & Morgan (*see also* Appendix).

PHYSALIS (Chinese-lantern plant)

When to plant: Indoors, 8 to 10 weeks before planting-out time in the spring; outdoors, in a seedframe, after the weather is warm and there is little danger of hard frost.

Temperature: 65°–75° F.

Special treatment: Seeds need light to sprout; sow on surface, and do not cover with planting medium.

Days to sprout: 15.

Light for seedlings: Half-day or more of sun, or start in a fluorescent-light garden (15–16 hours of light daily).

Maturity time: May give some bloom in the first season if started early indoors.

Comments: This perennial is easy to grow and is cultivated mostly for the colorful seedpods, pods which are effective in the garden as well as for dried arrangements. The seeds are listed by Butcher and Park (*see also* Appendix).

PHYSOSTEGIA (obedience plant)

When to plant: Outdoors, in a seedframe, in spring or sum-

mer; indoors, 6 to 10 weeks before planting-out time in the spring.

Temperature: 70° F.

Special treatment: None required.

Days to sprout: 15.

Light for seedlings: Half-day or more of sun, or start in a fluorescent-light garden (15–16 hours of light daily).

Maturity time: First blooms in the second season.

Comments: This hardy perennial is easy to grow; it is excellent for garden effect and for use as a cut flower. The seeds are listed by Butcher, Park, and Thompson & Morgan (*see also* Appendix).

PINCUSHION FLOWER: *See Scabiosa*

PINK: *See Dianthus*

PLANTAIN-LILY: *See Hosta*

PLATYCODON (balloonflower)

When to plant: Outdoors, in a seedframe, in spring or summer; indoors, 8 to 10 weeks before planting-out time in the spring.

Temperature: 70° F.

Special treatment: Seeds need light in order to sprout; sow on the surface, and do not cover with planting medium.

Days to sprout: 10.

Light for seedlings: Half-day or more of sun, or start in a fluorescent-light garden (15–16 hours of light daily).

Maturity time: Some blooms in the first season if seeds are started early indoors, as described above.

Comments: *Platycodon* is one of the best of all hardy perennials for planting in a flower garden or rockery. Balloonlike buds open into five-pointed, star- or bell-shaped flowers. The best known *Platycodon* has blue flowers, but it is also available in white and pink varieties. The seeds are listed by Butcher, Park, and Thompson & Morgan (*see also* Appendix).

POLEMONIUM (Jacob's-ladder, Greek valerian)

When to plant: Outdoors, in a seedframe, in late fall or early spring.
Temperature: 65° F.
Special treatment: None required.
Days to sprout: 20.
Light for seedlings: Half-day or more of sun.
Maturity time: First blooms in the second season.
Comments: Easy to grow from seeds, this hardy perennial deserves space in most flower gardens. The seeds are listed by Butcher, Park, and Thompson & Morgan (*see also* Appendix).

POTENTILLA (cinquefoil)

When to plant: Outdoors, in a seedframe, in late fall or early spring.
Temperature: 65° F.
Special treatment: None required.
Days to sprout: 15.
Light for seedlings: Half-day or more of sun.
Maturity time: First blooms in the second season.
Comments: Once established, *Potentilla* is a tough perennial that can be depended on for flowers over a long season. The seeds are listed by Butcher, Park, and Thompson & Morgan (*see also* Appendix).

PRIMROSE: *See Primula*

PRIMULA (primrose)

When to plant: Outdoors, in a seedframe, in late fall or early spring; indoors, in spring following refrigerator treatment (see below).
Temperature: 70° F.
Special treatment: Seeds may require freezing in order to sprout; to start indoors, first place planting in freezer for 2 to 4 months. Primrose seeds also need light in order to sprout; sow on surface, and do not cover with planting medium.
Days to sprout: 25 or more.
Light for seedlings: Half-day of sun, or start in a fluorescent-light garden (15–16 hours of light daily).
Maturity time: First blooms in the second season.
Comments: Primroses are among the most rewarding of all hardy perennials to grow from seeds. In the garden they need shade in the summer and soil that is rich in humus, moist, and well-drained. Seeds are listed by Burpee, Butcher, Park, Suttons, and Thompson & Morgan (*see also* Appendix).

PURPLE CONEFLOWER: *See Echinacea*

PYRETHRUM: *See Chrysanthemum*

RED HOT POKER: *See Kniphofia*

ROCK-CRESS: *See Arabis*

ROCK-FOIL: *See Saxifraga*

ROSEMALLOW: *See Hibiscus*

RUDBECKIA (coneflower)

When to plant: Outdoors, in a seedframe, in spring or sum-

mer; indoors, 8 to 10 weeks before planting-out time in the spring.

Temperature: 70° F.

Special treatment: None required.

Days to sprout: 20.

Light for seedlings: Half-day or more of sun, or start in a fluorescent-light garden (15–16 hours of light daily).

Maturity time: First blooms in the second season.

Comments: This hardy perennial is easy to grow from seeds and is cultivated both for garden effect and for use as a cut flower. The seeds are listed by Burpee, Butcher, Suttons, Park, and Thompson & Morgan (*see also* Appendix).

SAGE: *See Salvia*

SAGINA (pearlwort; Irish moss)

When to plant: Outdoors, in a seedframe, in early spring while the soil is cool.

Temperature: 65° F.

Special treatment: None required.

Days to sprout: 15.

Light for seedlings: Half-day of sun.

Maturity time: Interesting, mosslike ground cover within a few months.

Comments: This relative of *Dianthus*, or pinks, is one of the few moss look alikes that will grow in sun. The seeds are listed by Park and Thompson & Morgan (*see also* Appendix).

ST.-JOHN'S-WORT: *See Hypericum*

SALVIA (sage)

When to plant: Outdoors, in a seedframe, in spring or sum-

mer; indoors, 8 to 10 weeks before planting-out time in the spring.

Temperature: 70° F.

Special treatment: None required.

Days to sprout: 15.

Light for seedlings: Half-day or more of sun, or start in a fluorescent-light garden (15–16 hours of light daily).

Maturity time: First blooms in the second season, although *Salvia* started early indoors, as described above, may bloom toward the end of summer in the first season.

Comments: Perennial *Salvia* is easy to grow from seed and is an outstanding plant for almost any flower garden that receives a half-day or more of sun. The seeds are listed by Butcher, Park, and Thompson & Morgan (*see also* Appendix).

SANDWORT: *See Arenaria*

SANTOLINA (lavender-cotton)

When to plant: Outdoors, in a seedframe, in spring or summer; indoors, 8 to 10 weeks before planting-out time in the spring.

Temperature: 70° F.

Special treatment: None required.

Days to sprout: 15.

Light for seedlings: Half-day or more of sun, or start in a fluorescent-light garden (15–16 hours of light daily).

Maturity time: Effective foliage plant by the second season.

Comments: Once established, *Santolina* tolerates drought. It is cultivated mostly as a gray or green, low-growing edger for flower beds and borders. The seeds are listed by Park (*see also* Appendix).

SAPONARIA (soapwort; bouncing bet)

When to plant: Outdoors, in a seedframe, in spring or sum-

mer; indoors, 8 to 10 weeks before planting-out time in the spring.

Temperature: 65°–75° F.

Special treatment: Continual darkness, until the seeds sprout, produces the most successful germination.

Days to sprout: 10.

Light for seedlings: Half-day or more of sun, or start in a fluorescent-light garden (15–16 hours of light daily).

Maturity time: First blooms in the second season.

Comments: *Saponaria* is extremely easy to grow; it tolerates drought. The seeds are listed by Butcher, Park, Suttons, and Thompson & Morgan (*see also* Appendix).

SAXIFRAGA (saxifrage; rock-foil)

When to plant: Outdoors, in a seedframe, in late fall or early spring; indoors, in spring following refrigerator treatment (see below).

Temperature: 70° F.

Special treatment: Seeds may require freezing in order to sprout; to start indoors, first place planting in freezer for 2 to 4 months.

Days to sprout: 15.

Light for seedlings: Up to a half-day of sun, or start in a fluorescent-light garden (15–16 hours of light daily).

Maturity: First blooms in the second season.

Comments: *Saxifraga* is a choice hardy perennial for growing in a sunny rock garden with perfectly drained soil. The seeds are listed by Butcher, Park, Suttons, and Thompson & Morgan (*see also* Appendix).

SAXIFRAGE: *See Saxifraga*

SCABIOSA (scabious; pincushion flower)

When to plant: Outdoors, in a seedframe, in spring or summer; indoors, 8 to 10 weeks before planting-out time in the spring.
Temperature: 65°–75° F.
Special treatment: None required.
Days to sprout: 10 to 18.
Light for seedlings: Half-day or more of sun, or start in a fluorescent-light garden (15–16 hours of light daily).
Maturity time: First blooms in the second season.
Comments: *Scabiosa* is easy to grow from seeds and is valuable especially as a cut flower. The seeds are listed by Burpee, Butcher, Park, Suttons, and Thompson & Morgan (*see also* Appendix).

SCABIOUS: *See Scabiosa*

SEA-HOLLY: *See Eryngium*

SEA PINK: *See Armeria*

SEDUM (stonecrop; live forever)

When to plant: Outdoors, in a seedframe, in spring or summer; indoors, 8 to 10 weeks before planting-out time in the spring.
Temperature: 70° F.
Special treatment: None required.
Days to sprout: 5.
Light for seedlings: Half-day or more of sun, or start in a fluorescent-light garden (15–16 hours of light daily).
Maturity time: Effective ground cover by the second season.
Comments: These popular, hardy perennial succulents are easy to grow in any site that is sunny and has well-drained

soil; they are very drought-tolerant. The seeds are listed by Park and Thompson & Morgan (*see also* Appendix).

SEMPERVIVUM (houseleek; live forever)

When to plant: Outdoors, in a seedframe, in spring or summer; indoors, 8 to 10 weeks before planting-out time in the spring.
Temperature: 70° F.
Special treatment: None required.
Days to sprout: 15.
Light for seedlings: Half-day or more of sun, or start in a fluorescent-light garden (15–16 hours of light daily).
Maturity time: Effective ground cover by the second season.
Comments: These hardy succulents are invariably easy to grow; they tolerate drought, but some moisture and fairly rich, well-drained soil produces the best growth. The seeds are listed by Butcher, Park, and Thompson & Morgan (*see also* Appendix).

SHASTA DAISY: *See Chrysanthemum*

SHOOTING-STAR: *See Dodecatheon*

SILENE (campion; catchfly)

When to plant: Outdoors, in a seedframe, in spring or summer; indoors, 8 to 10 weeks before planting-out time in the spring.
Temperature: 70° F.
Special treatment: None required.
Days to sprout: 5.
Light for seedlings: Half-day or more of sun, or start in a fluorescent-light garden (15–16 hours of light daily).

Maturity time: First blooms in the second season; if started early indoors, as described above, may give some bloom toward the end of the first summer.

Comments: *Silene* is very easy to grow from seeds and is excellent for garden effect as well as a cut flower. The seeds are listed by Butcher, Park, Suttons, and Thompson & Morgan (*see also* Appendix).

SNEEZEWEED: *See Helenium*

SNOW-IN-SUMMER: *See Cerastium*

SOAPWORT: *See Saponaria*

SPANISH-DAGGER: *See Yucca*

SPEEDWELL: *See Veronica*

SPIDERWORT: *See Tradescantia*

STACHYS (lamb's-ears; betony)

When to plant: Outdoors, in a seedframe, in spring or summer; indoors, 8 to 10 weeks before planting-out time in the spring.

Temperature: 70° F.

Special treatment: None required.

Days to sprout: 15.

Light for seedlings: Half-day or more of sun, or start in a fluorescent-light garden (15–16 hours of light daily).

Maturity time: Effective silver-leaved ground cover by the second season, which is also when the first of the carmine red flowers appear.

Comments: *Stachys* is an excellent edger or ground cover in any site that is sunny with well-drained soil; it tolerates drought. The seeds are listed by Butcher, Park, and Thompson & Morgan (*see also* Appendix).

STOKES' ASTER: *See Stokesia*

STOKESIA (Stokes' aster)

When to plant: Outdoors, in a seedframe, in spring or summer; indoors, 8 to 10 weeks before planting-out time in the spring.
Temperature: 70° F.
Special treatment: None required.
Days to sprout: 20.
Light for seedlings: Half-day or more of sun, or start in a fluorescent-light garden (15–16 hours of light daily).
Maturity time: First blooms in second season; if started early indoors, as described above, some flowers may appear toward the end of the first summer.
Comments: Easy to grow from seeds, this perennial is excellent for garden effect as well as for cut flowers to use in bouquets. The seeds are listed by Park and Thompson & Morgan (*see also* Appendix).

STONECROP: *See Sedum*

SWEET WILLIAM: *See Dianthus*

TEUCRIUM (germander)

When to plant: Outdoors, in a seedframe, in spring or summer; indoors, 8 to 10 weeks before planting-out time in the spring.
Temperature: 70° F.

Special treatment: None required.
Days to sprout: 30.
Light for seedlings: Half-day or more of sun, or start in a fluorescent-light garden (15–16 hours of light daily).
Maturity time: Effective as an edger or very low hedge for flower beds and borders by the second season.
Comments: *Teucrium* needs moist, rich, well-drained soil. The seeds are listed by Park and Thompson & Morgan (*see also* Appendix).

THALICTRUM (meadow-rue)

When to plant: Outdoors, in a seedframe, in late fall or early spring.
Temperature: 65° F.
Special treatment: None required.
Days to sprout: 15.
Light for seedlings: Half-day or more of sun.
Maturity time: First blooms in the second season.
Comments: *Thalictrum* makes a beautiful plant and does best in moist soil that is rich in humus and well-drained. The seeds are listed by Butcher, Park, and Thompson & Morgan (*see also* Appendix).

THERMOPSIS

When to plant: Outdoors, in a seedframe, in spring or summer; indoors, 8 to 10 weeks before planting-out time in the spring.
Temperature: 70° F.
Special treatment: None required.
Days to sprout: 15.
Light for seedlings: Half-day or more of sun, or start in a fluorescent-light garden (15–16 hours of light daily).
Maturity time: First blooms in the second season.

Comments: *Thermopsis* is easy to grow from seeds and makes an outstanding flowering plant in a border. The seeds are listed by Park and Thompson & Morgan (*see also* Appendix).

THRIFT: *See Armeria*

TICKSEED: *See Coreopsis*

TORCH-LILY: *See Kniphofia*

TRADESCANTIA (spiderwort)

When to plant: Outdoors, in a seedframe, in late fall or early spring.
Temperature: 70° F.
Special treatment: None required.
Days to sprout: 10.
Light for seedlings: Half-day or more of sun.
Maturity time: First blooms in the second season.
Comments: *Tradescantia* does best in moist, well-drained, humuslike soil; however, once established, it tolerates considerable drought. It is available in both blue- and white-flowered varieties. The seeds are listed by Park and Thompson & Morgan (*see also* Appendix).

TRILLIUM (wake-robin)

When to plant: Outdoors, in a seedframe, in late fall or early winter; indoors, in early spring following refrigerator treatment (see below).
Temperature: 70° F.
Special treatment: Seeds require freezing in order to sprout; to start indoors, first place planting in freezer for 2 to 4 months.
Days to sprout: 50 or more.

Light for seedlings: Up to a half-day of sun, or start in a fluorescent-light garden (15–16 hours of light daily).
Maturity time: First blooms in 3 or 4 years.
Comments: *Trillium* is not easy to grow from seeds, but it is among the choicest of all early spring-flowering hardy perennials for a shaded, moist site with rich, well-drained soil. The seeds are listed by Park and Thompson & Morgan (*see also* Appendix).

TRITOMA: *See Kniphofia*

TROLLIUS (globeflower)

When to plant: Outdoors, in a seedframe, in late fall or early spring.
Temperature: 65° F.
Special treatment: None required.
Days to sprout: 50 or more.
Light for seedlings: Half-day or more of sun.
Maturity time: First blooms in the second season.
Comments: Although not easy to grow from seeds, *Trollius* is a beautiful, hardy perennial flower in a garden with rich, moist, well-drained soil. The seeds are listed by Butcher, Park, and Thompson & Morgan (*see also* Appendix).

TROUT-LILY: *See Erythronium*

VERONICA (speedwell)

When to plant: Outdoors, in a seedframe, in spring or summer; indoors, 8 to 10 weeks before planting-out time in the spring.
Temperature: 70° F.
Special treatment: None required.

Days to sprout: 15.
Light for seedlings: Half-day or more of sun, or start in a fluorescent-light garden (15–16 hours of light daily).
Maturity time: First blooms in the second season.
Comments: *Veronica* is easy to grow from seed, and it varies in size from ground carpeters to spire-form flowers several feet tall. The flowers are excellent for garden effect as well as for cutting. The seeds are listed by Butcher, Park, Suttons, and Thompson & Morgan (*see also* Appendix).

VIOLA (violet)

When to plant: Outdoors, in a seedframe, in late fall or early spring.
Temperature: 65° F.
Special treatment: None required.
Days to sprout: 10 or more.
Light for seedlings: Up to a half-day of sun.
Maturity time: First blooms in the second season.
Comments: True violets are ideal for use in nearly wild gardens, partly shaded rockeries, or as ground covers. Other species of *Viola* include pansy and Johnny-jump-up, both of which are covered in Chapter 4. Violet seeds are listed by Park and Thompson & Morgan (*see also* Appendix).

VIOLET: *See Viola*

WAHLENBERGIA

When to plant: Outdoors, in a seedframe, in spring or summer; indoors, 8 to 10 weeks before planting-out time in the spring.
Temperature: 70° F.
Special treatment: None required.
Days to sprout: 10.

Light for seedlings: Half-day or more of sun, or start in a fluorescent-light garden (15–16 hours of light daily).
Maturity time: First blooms in the second season.
Comments: Easy to grow from seeds, *Wahlenbergia* is an excellent cut flower. The seeds are listed by Park (*see also* Appendix).

WAKE-ROBIN: *See Trillium*

WELSH-POPPY: *See Meconopsis*

WINDFLOWER: *See Anemone*

WORMWOOD: *See Artemisia*

YARROW: *See Achillea*

YUCCA (Adam's needle; Spanish-dagger)

When to plant: Outdoors, in a seedframe, in late fall or early winter; indoors, in early spring following refrigerator treatment (see below).
Temperature: 70° F.
Special treatment: Seeds require freezing in order to sprout; to start indoors, first place planting in freezer for 2 to 4 months.
Days to sprout: 50 or more.
Light for seedlings: Half-day or more of sun, or start in a fluorescent-light garden (15–16 hours of light daily).
Maturity time: First blooms in 3 or 4 years.
Comments: This hardy perennial succulent is not easy to grow from seeds, but once established it makes a stunning effect in a sunny flower garden. It will tolerate drought better than almost any other flowering perennial. The seeds are listed by Park and Thompson & Morgan (*see also* Appendix).

6

Bulb Flowers to Grow from Seeds

Theoretically, all bulb flowers may be propagated from seeds. However, in this chapter I have chosen to include only those available from retail seed firms. If you acquire seeds of bulbs not discussed here, follow these general guidelines: if the seeds are from a bulb that grew originally in a temperate climate—tulip, daffodil, or hyacinth for example—treat as *Eranthis*, which is included in this chapter; if the seeds are from a bulb that hails from a tropical or subtropical climate—crinum lily for example—treat as *Agapanthus* or *Hedychium*, both of which are also included in this chapter.

ACHIMENES (magic flower)

When to plant: Winter or spring for bloom the following summer or fall.
Temperature: 70° F.
Special treatment: The seeds are exceedingly small; sow on the surface of the planting medium, and take care to keep it evenly moist at all times.
Days to sprout: 15.
Light for seedlings: 2 to 4 hours of sun (less in summer), or grow in a fluorescent-light garden (15–16 hours of light daily).
Maturity time: First blooms in 5 to 6 months. *Achimenes* forms scaly rhizomes in the soil; these multiply and can be kept over from year to year by storing them during fall and winter in barely damp vermiculite at 50–65° F.

Comments: When *Achimenes* seedlings have 3 or 4 sets of leaves, pinch out the tip growth of each in order to encourage compact, well-branched plants. *Achimenes* are primarily summer-flowering and are among the showiest of all hanging basket plants for partly shaded, outdoor gardens or for an east- or west-facing window inside. Seeds are available from Park (*see also* Appendix).

ACONITE, WINTER: *See Eranthis*

AGAPANTHUS (lily-of-the-Nile)

When to plant: Anytime.
Temperature: 70° F.
Special treatment: None required.
Days to sprout: 20 or more.
Light for seedlings: Half-day of sun, or grow in a fluorescent-light garden (15–16 hours of light daily).
Maturity time: First flowers appear in 3 to 5 years above long, strap-shaped leaves.
Comments: *Agapanthus* needs a growing medium that is constantly moist; otherwise, older leaves die rapidly and prematurely. This bulb is an outstanding landscape plant in outdoor gardens where frost is rarely experienced. Seeds of the blue-flowered *A. africanus* are listed by Park (*see also* Appendix).

ALLIUM (ornamental onion)

When to plant: Early spring or late fall.
Temperature: 50–70° F.
Special treatment: Cover seeds to the depth of their own thickness with growing medium and moisten well; place in bottom of refrigerator or in a frame outdoors for at least a month of chilling or even freezing temperatures; then bring to light and warmth.

Days to sprout: 15, following special treatment (*see above*).
Light for seedlings: Half-day or more of sun, or grow in a fluorescent-light garden (15–16 hours of light daily).
Maturity time: First flowers appear in 2 to 4 years.
Comments: These relatives of the onions we eat come in a variety of sizes, from 1 to 4 feet tall, and offer blooms of many colors over a long season. They are prized hardy perennials for a sunny border or rock garden. Seeds are listed by Park and Thompson & Morgan (*see also* Appendix).

AMARYLLIS: *See Hippeastrum*

ANEMONE

When to plant: In spring for bloom the following late winter and spring.
Temperature: 65° F.
Special treatment: None required.
Days to sprout: 15.
Light for seedlings: Half-day or more of sun, or grow in a fluorescent-light garden (15–16 hours of light daily).
Maturity time: First flowers in 10 to 12 months.
Comments: Seedlings of the showy, poppy-flowered *Anemones* form tuberous roots, which may be kept from year to year. During the winter, *Anemones* need to be cool (45°–55° F.). They also need as much sun as possible and fresh, moist air that circulates freely. In climates where winter seldom brings freezing temperatures, grow them outdoors; elsewhere these *Anemones*, usually called St. Brigid or Monarch De Caen, are choice winter- and spring-flowering plants for a cool greenhouse. Seeds are listed by Park and Thompson & Morgan (*see also* Appendix). For other *Anemones* to grow from seeds, *see also* Chapter 5.

BEGONIA, TUBEROUS

When to plant: Sow seeds indoors 12 to 16 weeks before

Tuberous-rooted begonias like these have tiny seeds that must be started indoors in controlled conditions. Allow about six months from planting to first blooms. Tubers which form in the soil can be kept year after year.
Courtesy Maynard Parker.

frost-free weather is expected outdoors.
Temperature: 65° F.
Special treatment: Seeds need light to sprout; sow on surface, and do not cover with planting medium.
Days to sprout: 15.
Light for seedlings: 2 hours of sun, bright north light, or a fluorescent-light garden (15–16 hours of light daily).
Maturity time: 16–20 weeks to first blooms.
Comments: *Begonia* seeds are dust-sized, but surprisingly easy to grow. Seedlings form tubers which can be stored over winter in barely damp vermiculite at 50°–65° F. Seeds of tuberous-rooted hybrids are listed by Antonelli, Burpee, Butcher, Park, Suttons, and Thompson & Morgan (*see also* Appendix).

BUTTERCUP: *See Ranunculus*

BUTTERFLY LILY: *See Hedychium*

CANNA

When to plant: January–March for bloom the following summer.
Temperature: 68°–78° F.
Special treatment: Soak seeds in water at room temperature for 48 hours; then use a nail file to chip each before sowing.
Days to sprout: 21–28.
Light for seedlings: Half-day or more of sun, or grow in a fluorescent-light garden (15–16 hours of light daily).
Maturity time: First flowers 16–18 weeks following germination.
Comments: *Canna* seedlings form tuberous roots that may be kept over winter in barely damp vermiculite at 50°–65° F. Grow them outdoors in warm weather for beautiful flowers and attractive foliage, either in a border or in containers (12

inches in diameter or larger). Seeds are listed by Butcher, Park, Suttons, and Thompson & Morgan (*see also* Appendix).

CLIVIA (Kaffir-lily)

When to plant: Spring is the best season.
Temperature: 70°–75° F.
Special treatment: None required.
Days to sprout: 30.
Light for seedlings: Half-day or more of sun, or grow in a fluorescent-light garden (15–16 hours of light daily).
Maturity time: 6 to 7 years before first blooms.
Comments: *Clivia* makes a handsome foliage plant with leathery, dark green, strap-shaped leaves that are arranged precisely in a fan. The umbels of flowers may be apricot, salmon, scarlet, or yellow. If pollinated, some will form fruits that are green at first, but then ripen to bright red. *Clivia* may be grown outdoors yearlong in frost-free climates; elsewhere grow it in a pot or tub. Seeds are listed by Park and Thompson & Morgan (*see also* Appendix).

CYCLAMEN (hardy)

When to plant: October–March.
Temperature: 60° F.
Special treatment: Sow seeds ¼ inch deep, and place in a cool (40°–50° F.) window, greenhouse, or frame outdoors.
Days to sprout: 50 or more.
Light for seedlings: Half-day of sun or grow in a fluorescent-light garden (15–16 hours of light daily).
Maturity time: Allow 2 years for first flowers. When seedlings have several leaves, transplant to moist, humus-rich soil in a partly shaded border or wildflower garden.
Comments: Seeds of *Cyclamen neapolitanum*, the best-known hardy species (as distinguished from the frost-sensitive florist

These hardy Cyclamen seedlings are about one year old. Started from seeds planted outdoors in late fall or early winter, they are now large enough to plant directly in the open garden.

types), are listed by Butcher, Park, Suttons, and Thompson & Morgan (*see also* Appendix); this species has silver-marbled, heart-shaped leaves and rose red flowers. Numerous other hardy *Cyclamen* seeds are listed by Park.

DAHLIA

When to plant: 8 to 12 weeks before warm, frost-free weather is expected outdoors.
Temperature: 70° F.
Special treatment: None required.
Days to sprout: 5.
Light for seedlings: Half-day or more of sun, or grow in a fluorescent-light garden (15–16 hours of light daily). Pinch out the tips frequently to encourage branching.
Maturity time: First flowers approximately 4 months after sowing.
Comments: *Dahlia* is extremely easy to grow from seeds. During the first season it forms tubers in the ground which may be dug in autumn, just before frost, and stored over winter at 50°–60° F. in barely damp vermiculite. Hybrid seeds of many beautiful strains are listed by Burpee, Butcher, Park, Suttons, and Thompson & Morgan (*see also* Appendix).

ERANTHIS (winter aconite)

When to plant: Early spring or late fall.
Temperature: 50–60° F.
Special treatment: Cover seeds to the depth of their own thickness with growing medium and moisten well; place in freezer or in a frame outdoors for at least a month of freezing temperatures; then bring plants to light and warmth.
Days to sprout: 30 or more.
Light for seedlings: Half-day or more of sun, or grow in a fluorescent-light garden (15–16 hours of light daily).

When Dahlia *seedlings have formed about four sets of leaves, pinch out the growing tips (as has been done to these) to promote compact, well-branched growth.*

Maturity time: First flowers appear in 2 or 3 years, in late winter or earliest spring.

Comments: This hardy little bulb grows less than 6 inches tall, forming a tuft of leaves topped by buttercup yellow flowers. Plant in clumps or drifts in a sunny border or rock garden, very much the same as *Crocus*. Seeds are listed by Park and Thompson & Morgan (*see also* Appendix).

EUCOMIS (pineapple-lily)

When to plant: Anytime.

Temperature: 70° F.

Special treatment: None required; easy to grow.

Days to sprout: 20.

Light for seedlings: A half-day of sun, or grow in a fluorescent-light garden (15–16 hours of light daily).

Maturity time: 3 to 4 years for the first flowers—greenish clusters at the top of a stalk, crowned by a rosette of leaves similar to a pineapple.

Comments: Seedlings form a bulb, which can be kept from year to year. However, *Eucomis* is a tender bulb; it should be treated as *Gladiolus* in climates where winter brings freezing temperatures. Seeds are listed by Park and Thompson & Morgan (*see also* Appendix).

FREESIA

When to plant: Winter for flowers the following winter.

Temperature: 65° F.

Special treatment: None required.

Days to sprout: 25.

Light for seedlings: Half-day or more of sun, or grow in a fluorescent-light garden (15–16 hours of light daily).

Maturity time: First flowers in 9 to 12 months.

Comments: *Freesia* seedlings form small corms (bulbs) in the soil. Do not allow them to dry out severely at any time, or they will become dormant and flowering will be delayed. During the winter heating season, *Freesia* needs temperatures on the cool side, especially at night (45°–55° F.). They also should have fresh, moist air that circulates freely. These plants are ideal for winter flowers in a cool greenhouse, or outdoors in frost-free climates. The fragrance of *Freesia* flowers is the favorite scent of many persons. Seeds of various hybrid strains are listed by Butcher, Park, Suttons, and Thompson & Morgan (*see also* Appendix).

GALTONIA (summer hyacinth)

When to plant: Spring or early summer.
Temperature: 65°–70° F.
Special treatment: None required.
Days to sprout: 15.
Light for seedlings: Half-day or more of sun, or grow in a fluorescent-light garden (15–16 hours of light daily).
Maturity time: 2 or 3 years for first flowers.
Comments: This bulb is hardy outdoors during the winter as far north as Philadelphia; elsewhere treat it as *Gladiolus*. It is prized for three-foot spikes of white, bell-shaped flowers in late summer. Seeds are listed by Park and Thompson & Morgan (*see also* Appendix).

GARLAND FLOWER: *See Hedychium*

GINGER-LILY: *See Hedychium*

GLADIOLUS

When to plant: Early spring while the soil is cool.
Temperature: 50°–60° F.

Special treatment: Sow seeds to the depth of their own thickness, in a protected frame outdoors, as soon as the soil can be worked in the spring.
Days to sprout: 20.
Light for seedlings: Half-day or more of sun.
Maturity time: First flowers in the second summer.
Comments: *Gladiolus* seedlings form a corm (bulb) in the ground during the first season. Just before the fall frost, dig corms up, clean and store them indoors until planting-out time the following spring. Seeds are listed by Park (*see also* Appendix).

GLORIOSA (climbing lily)

When to plant: Anytime.
Temperature: 70° F.
Special treatment: None required.
Days to sprout: 30.
Light for seedlings: A half-day or more of sun, or grow in a fluorescent-light garden (15–16 hours of light daily).
Maturity time: 18 months to 2 years for first flowers, which are usually yellow and red.
Comments: These exotic flowers need constant warmth and high humidity in air that circulates freely. Seedlings form thick, white icicle, radishlike roots. These may be kept from year to year, stored during the dormant season in barely damp vermiculite at 60°–70° F. *Gloriosa* may be treated as *Gladiolus* in climates where freezing temperatures occur in winter. Provide a trellis or other support. Seeds are listed by Butcher, Park, and Thompson & Morgan (*see also* Appendix).

HEDYCHIUM (ginger-lily; butterfly lily; garland flower)

When to plant: Winter or spring.

Temperature: 68°–78° F.
Special treatment: None required.
Days to sprout: 25.
Light for seedlings: Half-day or more of sun, or grow in a fluorescent-light garden (15–16 hours of light daily).
Maturity time: First flowers in 2 to 3 years.
Comments: *Hedychium* may be grown outdoors all year in frost-free climates. Elsewhere treat it as *Canna*, storing the tuberous roots over winter in barely damp vermiculite at 50°–65° F. *Hedychium coronarium* bears fragrant white flowers in warm weather and is widely planted in outdoor gardens of the deep South. Seeds of *H. coccineum*, a red-flowered form, are listed by Park (*see also* Appendix).

HIPPEASTRUM (amaryllis)

When to plant: Anytime.
Temperature: 70° F.
Special treatment: The seeds are papery and black, some as large as a nickel. They may be sown flat on the surface or sideways in narrow drills opened in the planting medium with a knife blade.
Days to sprout: 30.
Light for seedlings: Half-day or more of sun, or grow in a fluorescent-light garden (15–16 hours of light daily).
Maturity time: First flowers in 2 to 3 years.
Comments: Amaryllis may be grown outdoors all year in frost-free climates; elsewhere grow them as house plants except in warm weather. Seeds of both Dutch and American hybrids—as well as other fascinating species such as the blue amaryllis (first blooms in 5 to 7 years) and the green amaryllis (first blooms in 4 to 5 years)—are listed by Park (*see also* Appendix). Key's hybrid is said to bloom in less than 2 years from seeds; the flowers may be pink, red, rose, or white. Such hybrid amaryllis seeds are also listed by Butcher, Suttons, and Thompson & Morgan (*see also* Appendix).

HYACINTH, SUMMER: *See Galtonia*

KAFFIR-LILY: *See Clivia*

LILIUM (lily)

When to plant: Late fall, in a seed frame outdoors, if species or hybrids of *auratum*, *martagon*, *giganteum*, or *speciosum rubrum*; winter indoors or spring outdoors if species or hybrids of *formosanum*, *regale*, *tenuifolium*, *longiflorum*, or *centifolium*.

Temperature: 60°–70° F.

Special treatment: Species or hybrids of *auratum*, *martagon*, *giganteum*, and *speciosum rubrum* need several months of chilling or freezing temperatures to break the dormancy of the seeds. Sow about ¼ inch deep in a frame outdoors in late fall; sprouting should occur at the onset of warm weather the following spring. An alternate treatment for these lilies is to mix the seeds with equal parts moist sphagnum peat moss and sand; place this mixture in a jar with a screw-on lid that has a few holes punched in it; keep it in the refrigerator until the seeds have begun to form little bulbs; then plant bulbs ½ inch deep in pots or flats indoors, or outdoors in a seed frame.

Days to sprout: 30 or more.

Light for seedlings: Half-day or more of sun, or grow in a fluorescent-light garden (15–16 hours of light daily).

Maturity time: If started indoors in ideal conditions the winter before, seedlings of species or hybrids of *formosanum*, *regale*, *tenuifolium*, *longiflorum*, and *centifolium* may give some bloom the first summer. Others may not give first bloom for several years.

Comments: Seeds of species and hybrid lilies are listed by Butcher, Park, and Thompson & Morgan (*see also* Appendix). You can also cross pollinate lilies growing in your own garden, harvest the seeds, and raise new hybrids.

LILY: *See Lilium*

LILY, CLIMBING: *See Gloriosa*

LILY-OF-THE-NILE: *See Agapanthus*

MAGIC FLOWER: *See Achimenes*

ONION, ORNAMENTAL: *See Allium*

PINEAPPLE-LILY: *See Eucomis*

RANUNCULUS (buttercup)

When to plant: Winter or spring.
Temperature: 68° F.
Special treatment: None required.
Days to sprout: 14–42.
Light for seedlings: Half-day or more of sun, or grow in a fluorescent-light garden (15–16 hours of light daily).
Maturity time: 12–18 months to first blooms. Seedlings form tuberous roots, which may be kept from year to year.
Comments: During the winter heating season, *Ranunculus* needs temperatures on the cool side, especially at night (45°–55° F.). It also should have fresh, moist air that circulates freely. Buttercups are ideal for winter–spring flowers in a cool greenhouse, or outdoors in frost-free climates. Seeds of hybrid strains are listed by Butcher, Park, Suttons, and Thompson & Morgan (*see also* Appendix).

SHELLFLOWER: *See Tigridia*

TIGER-FLOWER: *See Tigridia*

TIGRIDIA (tiger-flower; shellflower)

When to plant: In the spring, as soon as the soil is warm.

Temperature: 50°–70° F.

Special treatment: Sow seeds to the depth of their own thickness in a protected frame outdoors.

Days to sprout: 30.

Light for seedlings: Half-day or more of sun.

Maturity time: First flowers toward the end of the first summer.

Comments: *Tigridia* seedlings form a corm (bulb) in the ground during the first season. Just before the fall frost, dig corms up; clean and store them indoors until planting-out time the following spring. During the first season it is best to let the seedlings grow undisturbed in the seedframe where they were planted. *Tigridia* seeds are listed by Burpee and Park (*see also* Appendix).

Appendix

SOURCES FOR FLOWER SEEDS AND SUPPLIES

Antonelli Brothers
2545 Capitola Road
Santa Cruz, CA 95010

W. Atlee Burpee Company
Warminster, PA 18974

Thomas Butcher, Ltd.
60 Wickham Road
Surrey C49 8AG
London
England

Nichols Herb and Rare Seeds
1190 North Pacific Highway
Albany, OR 97321

Park Seed
Greenwood, SC 29647

Sunnyslope Gardens
9638 Huntington Drive
San Gabriel, CA 91775

Suttons Seeds
161 New Bond Street
London W 1
England

Thompson and Morgan, Inc.
P.O. Box 24
401 Kennedy Boulevard
Somerdale, NJ 08083

PLANT SOCIETIES TO SERVE SPECIAL INTERESTS

The American Daffodil Society, Inc.
89 Chichester Road
New Canaan, CT 06840

The American Dahlia Society
163 Grant Street
Dover, NJ 07801

American Fern Society
Biological Sciences Group
University of Connecticut
Storrs, CT 06268

The American Fuchsia Society
Hall of Flowers

Golden Gate Park
San Francisco, CA 94122

American Gourd Society
P.O. Box 274
Mt. Gilead, OH 43338

American Hemerocallis Society
Signal Mountain, TN 37377

The American Hosta Society
114 The Fairway
Albert Lea, MN 56007

The American Penstemon Society
1547 Monroe Street
Red Bluff, CA 96080

American Peony Society
250 Interlachen Road
Hopkins, MN 55343

The American Plant Life Society
The American Amaryllis Society Group
Box 150
La Jolla, CA 92037

The American Primrose Society
7100 South West 209th
Beaverton, OR 97005

American Rhododendron Society
2232 N. E. 78th Avenue
Portland, OR 97213

American Rock Garden Society
90 Pierpont Road
Waterbury, CT 06705

California Native Plant Society
Suite 317
2490 Channing Way
Berkeley, CA 94704

Cymbidium Society of
   America, Inc.
6787 Worsham Drive
Whittier, CA 90602

Median Iris Society
10 South Franklin Circle
Littleton, CO 80121

National Chrysanthemum
   Society Inc., USA
394 Central Avenue
Mountainside, NJ 07092

New England Wild Flower
   Society, Inc.
Hemenway Road
Framingham, MA 01701

North American Gladiolus
   Council
30 Highland Place
Peru, IN 46970

Reblooming Iris Society
903 Tyler Avenue
Radford, VA 24141

Spuria Iris Society
Route 2, Box 83
Purcell, OK 73080

The Society of Japanese Irises
17225 McKenzie Highway
Route 2
Springfield, OR 97477

Society for Louisiana Irises
Box 175
University of SW Louisiana
Lafayette, LA 70501

The Society for Siberian Irises
South Harpswell, ME 07049

Index